Thinking Local
in Tompkins County

Articles from TCLocal.org

Volume 2

First printing December 2011

Pinax Publishing
Ithaca, New York
ISBN 978-0-615-57925-2

Introduction

TCLocal is a group of people living in and around Tompkins County, New York, dedicated to planning for a future of increasingly expensive energy, beginning with a crisis in the supply of liquid fuels that a number of analysts see coming by the end of this decade. Our proposed response is two-fold: *relocalization*—the rebuilding of the local sources of food and manufactured goods that have been lost over the last half century of globalization—and an emphasis on increased self-reliance at both the personal and organizational level.

Our approach to planning for relocalization has been to research and analyze the more important challenges facing us here in Tompkins County and then to publish articles reflecting our findings on our web site, TCLocal.org, in a format that promotes community review and response.

The first TCLocal.org article appeared in January 2008, and by mid-2009 enough had been published online for a print collection, *Thinking Local in Tompkins County, Vol. 1*. Through the generosity of an anonymous donor, we printed 500 copies, distributing them to most of the elected and appointed municipal officials and department heads in the County.

Funded by a grant from the Park Foundation, we have now gathered together the articles published at TCLocal.org over the succeeding two years into this second volume. These articles provide a fairly comprehensive introduction to the liquid fuels problem and possible local responses in the key areas of food production, heating, and health care.

This collection leads off with our outlook for the availability of liquid fuels over the coming decade and the impact it will have on our daily existence. A review of two Cornell projects that investigated regional food production and the concept of a "foodshed" then serves as prelude to the centerpiece of this collection, a vision of Tompkins County food production that originally ap-

peared online as a six-part series. The food production theme continues with an article on raising chickens in our area, transitioning in the two articles that follow to the related concepts of food security and health care. Three more articles provide an in-depth look at local biomass production for heat and power before the collection ends with a call for monetary investment in local production via "slow money."

It should be understood that these pieces capture a view of the immediate future based on the best analyses available in 2009–2010, all of which predicted constraints in the supply of liquid fuels beginning by 2020 (see pages 8–9). But ever since we started TCLocal in 2005, it has been a question whether a financial system built on bad debt and continuing growth will hold out long enough to maintain the consumption of energy that would actually run into physical constraints on the rate of production.

That limits on energy are in effect limits to growth would suggest a future of economic crashes and recoveries, as explained in the article "Outlook for Liquid Fuels" beginning on page 2 of this collection. It now appears possible, however, that due to financial chicanery, demand may not even recover enough to give the illusion of real economic recovery. If so, we may not experience the supply-side constraints associated with "peak oil" for some time, and the visible effects of resource depletion may fade from view in the context of general economic decline while remaining an invisible ceiling on growth.

However hazy the future may be, some basics can be counted on for planning purposes. Barring the invention of a completely new source of energy, two outcomes appear certain: in the future we will, collectively, have less material wealth (likely manifested as reduced purchasing power), and we will, collectively, have less mobility (which will mean spending a lot more time close to home). The big question is whether we will also have less happiness. The history of the past says that it does not have to be that way; in terms of the things that actually make people happy, we were, on average, better off a century ago than we are today. But past history also says that things can be a lot worse. The task before us is to prepare as graceful a transition as possible to a different and possibly better way of life. The articles in this volume represent a portion of TCLocal's contribution to that effort.

Jon Bosak, TCLocal Editor
Ithaca, December 2011

O

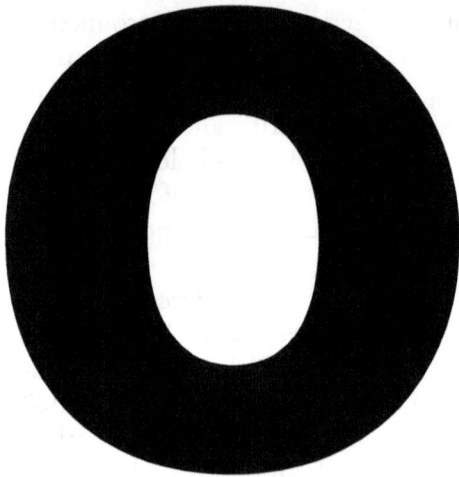

Outlook for Liquid Fuels, 2010-2020

By Jon Bosak (September 2010)

As our regular readers know, we here at TCLocal are engaged in a long-term effort to help develop local responses to energy descent—the condition of decreasingly available energy. The near-term manifestation of energy descent is high liquid fuel prices caused by a leveling and decline in global oil production, and it has been part of our job over the five years since we started TCLocal to keep an eye on the liquid fuel outlook and periodically advise the residents of Tompkins County (in particular, local policy makers) on what we're finding.

The September 2010 overview that follows was based on a fresh analysis of the current situation conducted by a team consisting of Jon Bosak, Bethany Schroeder, Karl North, and Tom Shelley. Not every detail will be agreed upon by every TCLocal contributor or even by every member of the research team, but it does represent in a general way a shared view of the outlook for liquid fuels over the next decade. Illustrations drawn from the Web are credited where the source is known and are reproduced here under the Fair Use provisions of copyright law.

THE LESSON OF DEEPWATER HORIZON

Predicting the price of oil is an extraordinarily difficult task even for petroleum experts, which we are not, and the effect of the economic downturn on the oil business has made reliable forecasting almost impossible in recent years. But in the summer of 2010, one thing, at least, became very clear: the easy oil is gone. That's not a future development; it's already here. No one pursues a course as risky, dangerous, and expensive as drilling four miles down into the Gulf of Mexico unless all the easier stuff is no longer available. It doesn't take a degree in petroleum engineering to see this.

U.S. Coast Guard

*(It is enlightening to understand some basic facts about oil extraction, though, so if you'd like to know more about that, we've prepared a brief tutorial on the subject at **http://ibiblio.org/tcrp/sidebars/extraction.html**)*

Here's a recently published official view of the future from the U.S. Department of Energy's Energy Information Agency (EIA).

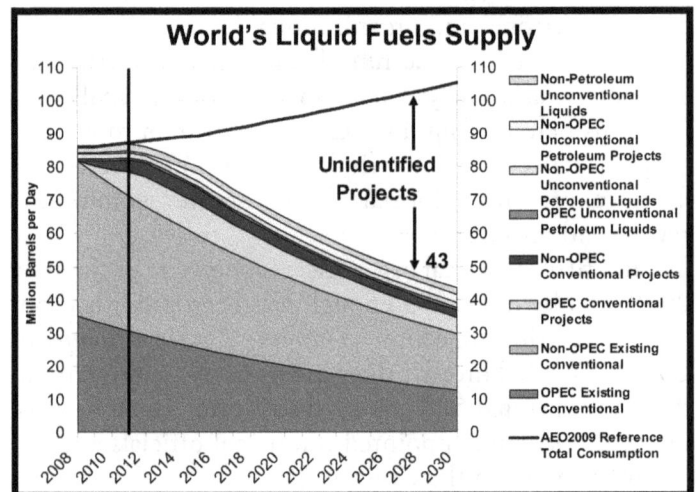

The top line in this remarkable graph is world demand for liquid fuels. Over the long term it *always* increases steadily due to population growth, if nothing else.

The colored areas show the global sources of liquid fuels, taking into account all currently known projects. As the graph makes clear, existing sources of conventional oil are already in steep decline, and unconventional sources can't keep up with that decline. The result is a growing gap between supply and demand beginning not long after 2012.

In the petroleum world, this has never happened before. Up until now, there has always been enough liquid fuel to meet demand, because it could be pumped out as fast as people had a use for it. A widening gap between supply and demand will eventually have an upward effect on prices beyond anything seen so far.

The label "Unidentified Projects" in the illustration acknowledges that no one really knows what sources can fill this widening gap between supply and demand. It is certain that no combination of currently foreseeable efforts can make up for the rate of decline in conventional oil production, and any new projects are certain to be much more expensive than those of the past.

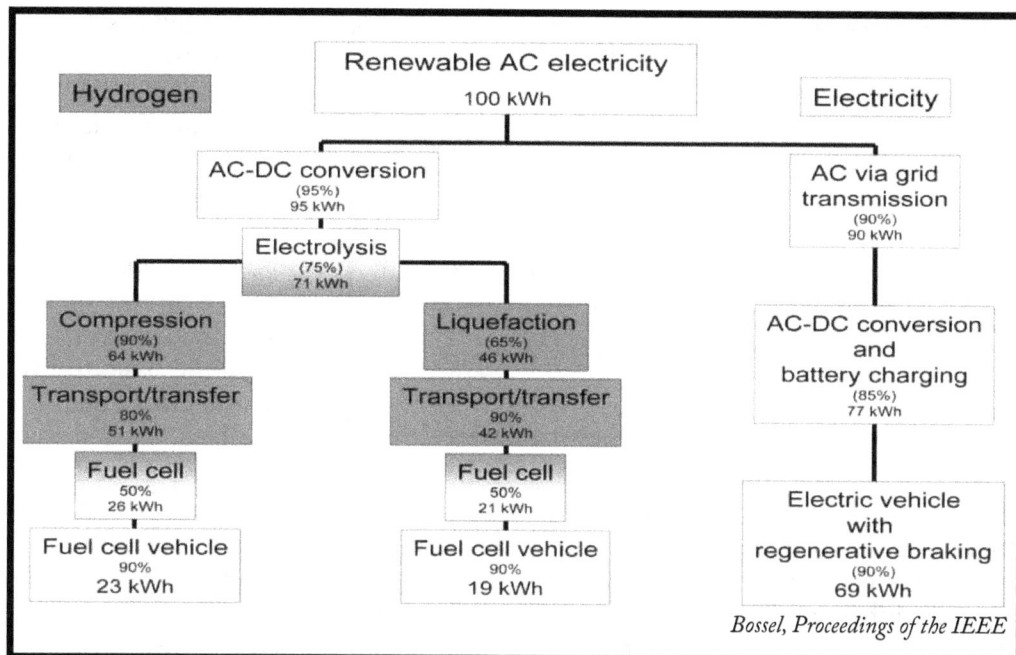

Bossel, *Proceedings of the IEEE*

RED HERRINGS AND DEAD ENDS

At this point, most readers will be thinking of their favorite solution to the energy problem. But within the ten-year period that we're discussing here, *there is no solution.* No current proposal will avert a near-term future of decreasingly less available liquid fuel.

This conclusion may come as a shock to anyone who's put their faith in technological fixes. We seem to have so many promising solutions to choose from; you'd think the problem was just getting them implemented. There certainly is a lot we could be doing that we aren't, but on examination it turns out that the proposed technological solutions to the coming oil crunch are at best wishful thinking and at worst border on the fraudulent.

A prime example of this latter category is the idea that we will replace current vehicles with ones fueled by hydrogen. The fact is that hydrogen is not a source of energy, it's just a way of storing energy, like batteries. And if we had the extra energy to store, we could distribute it much more easily by building out the existing electric grid—and much more efficiently, too. As shown in the figure above, the pure electric approach delivers three times the power to the road from a given input of electricity than the hydrogen-based approach.

Remember GM's relentless promotion of hydrogen cars? The last serious publicity the company put into this was in 2006. It's now obvious that this was all just a PR campaign designed to reassure consumers that GM was working toward a transition away from fossil fuels. That role is now being played by electric cars. This is an improvement, but unfortunately not a solution.

The proposal to solve the liquid fuels problem by transitioning to electricity is one of a large class of putative solutions that make some technical sense but just don't comprehend the scale of the problem. It's clear that widespread conversion to electric vehicles will require some kind of addition to our generating capacity, but few people appreciate the size of the change. The fact is that most people have no idea how much energy we're consuming to move as many vehicles around as we do.

Let's do a little back-of-the-envelope calculation here. According to the EIA, total U.S. petroleum consumption in 2007 was 20,680,000 barrels per day, and 70 percent of it went to transportation. A barrel of oil represents 1700 kWh of energy. Do the arithmetic and you'll find that transportation in the U.S. uses about 9.0 billion MWh/year of energy from petroleum. By comparison, according to the EIA, total U.S. electrical output in 2007 was about 4.2 billion MWh. In other words, the amount of energy represented by the fuel we're using in vehicles is more than twice as much as the total amount of energy represented by the electricity we're producing each year. Speaking very roughly, therefore, a proposal to replace half of our vehicle fleet with electric versions amounts to a proposal to double the size of our entire electric generating and distribution system, which includes doubling the amount of fuel consumed (chiefly coal). It is safe to assume that we will not see this happening in the next ten years, if ever.

Proposals that rely on solar, wind, or nuclear to provide the missing electricity demonstrate a similar failure to understand the scale of the problem. The following diagram illustrates this point.

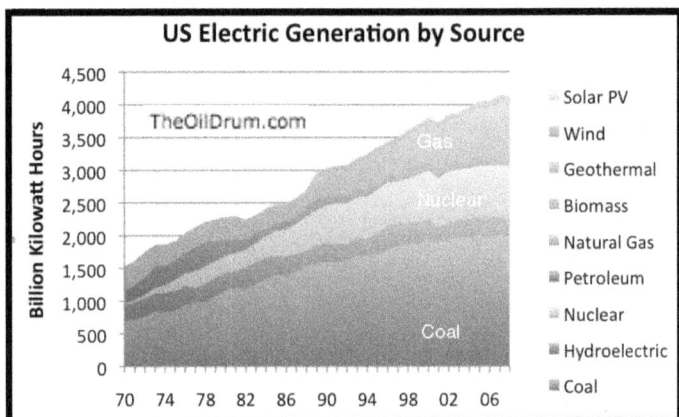

US Electric Generation by Source

TheOilDrum.com

Legend: Solar PV, Wind, Geothermal, Biomass, Natural Gas, Petroleum, Nuclear, Hydroelectric, Coal

Count off the sources by working up from the bottom of the graph and you'll begin to understand what a tiny proportion of our electrical generating capacity is due to wind, solar, geothermal, and biomass; their contribution is barely visible. Electricity from nuclear is much greater, of course, but the cost and planning horizon of nuclear projects means that any sizable expansion of nuclear capacity would lie many years in the future. Aside from the inability of these sources to provide liquid fuel, no believable expansion scenario envisions any combination of them being able to fill more than a fraction of the energy gap that's opening up due to the decline of conventional oil.

A third class of solutions *would* actually solve the liquid fuel problem, but only for a little while, and at an enormous cost in other resources. Hydrofracking for natural gas in our local Marcellus shale is an example of this category of solutions: we get a temporary shot of fossil fuel at the cost of our farms and our drinking water, and at the end of the process we're left back where we started but with permanent damage to our environment.

Another technology in this category is oil from "tar sands" and "oil shales," production of which uses phenomenally large amounts of water and is even more destructive to the environment than hydrofracking.

OUR ALTERNATIVE FUEL DIVISION HAS FOUND A WAY TO TURN FRESH WATER INTO FUEL!

WOULDN'T THAT TURN THE WORLD INTO AN UNINHABITABLE WASTELAND IN THE LONG RUN?

NOT IF SOMEONE FINDS A WAY TO TURN OIL INTO WATER.

© Scott Adams, Inc./Dist. by UFS, Inc.

Tar sands are also representative of a class of good-looking production technologies that don't yield significantly more energy than they use but simply substitute one source of energy for another, in this case, massive amounts of natural gas to heat the "tar" (bitumen). Another example of an energy "source" that doesn't actually

deliver significantly more energy than it consumes is liquid fuel from biomass, such as ethanol from corn.

To sum up, then: some of these alternatives—in particular, the development of solar and wind power—really are worth pursuing, but none of the current proposals can change the history of the next decade or so, either because they are not solutions at all or because it is physically impossible to increase production from alternative sources quickly enough to have a meaningful impact in that period of time. The only thing that could change the basic reality would be a massive, all-out effort to replace liquid fuels with substitutes from coal or natural gas.

Large-scale production of liquid fuels from coal has only been accomplished twice in history, once by the Nazi government in Germany and once by the apartheid regime in South Africa; the synthetic fuel is of excellent quality, but the technology is brutally expensive and therefore instituted only as a last resort. And of course large-scale coal-to-liquids would just delay the inflection point without really changing anything, because coal and natural gas are themselves finite resources that are closer to their own peaks than most people realize.

Coal-to-liquids shares one more flaw with most of the other proposed solutions: we're out of time. A 2005 study commissioned by the U.S. Department of Energy concluded that widespread disruption to our economic system from peak oil could be averted by nothing less than a WW2-level national mobilization effort to implement coal-to-liquids starting *at least* a decade ahead of the peak—and we don't have that kind of time left.

TIMING THE GAP

The inevitability of a coming liquid fuel price crisis caused by failure of oil production to meet increasing demand is much easier to establish than the precise timing of that crisis. But several independent studies have recently arrived from different directions at approximately the same conclusion.

In "The Status of Conventional World Oil Reserves," published recently in the journal *Energy Policy*, researchers Owen, Interwildi, and King conducted an in-depth survey of all currently available information regarding oil production and petroleum reserves, with special attention to the reliability of reporting in the OPEC countries. Their conclusion:

Supply and demand is likely to diverge **between 2010 and 2015,** unless demand falls in parallel with supply constrained induced recession.

Note the "unless"; we'll return to that shortly.

In the article "Forecasting World Crude Oil Production Using Multicyclic Hubbert Model," published last

April in *Energy & Fuels 2010,* a team from Kuwait University (Nashawi, Malallah, and Al-Bisharah) performed an in-depth mathematical analysis of the 47 leading oil-producing countries. While based on a methodology completely different from that used by Interwildi *et al.,* their findings are strikingly similar:

> World oil reserves are being depleted at an annual rate of 2.1%.... World production is **estimated to peak in 2014....**

A third independent study is notable for its source: the United States Joint Command (that is, the U.S. military establishment). Their official public assessment of the current situation, published last February in *Joint Operating Environment 2010,* is short on detail but very clear:

> **By 2012,** surplus oil production capacity could entirely disappear, and **as early as 2015,** the shortfall in output could reach nearly 10 MBD.

Ten million barrels per day (MBD) is about 12 percent of current global oil production. A shortfall of that magnitude would have an effect on fuel prices that's difficult to fully imagine.

The mainstream business press has until recently been notably dismissive of such estimates, regardless of the credibility of their sources (how can you dismiss the entire U.S. military?). But in September 2010, *Forbes,* which bills itself as the "capitalist tool," broke the wall of denial in an interview with respected oil analyst and oil industry veteran Charles Maxwell (nicknamed "the Dean of Oil Analysts"). Maxwell said:

> A bind is clearly coming. We think that the peak in production will actually occur in the period 2015 to 2020. And if I had to pick a particular year, I might use 2017 or 2018. That would suggest that **around 2015, we will hit a near-plateau of production around the world,** and we will hold it for maybe four or five years. On the other side of that plateau, production will begin slowly moving down. **By 2020, we should be headed in a downward direction** for oil output in the world each year instead of an upward direction, as we are today.

As might be expected, the estimate in *Forbes* is the most conservative of the forecasts quoted here, but even it clearly sees a fundamental change in the liquid fuels supply before the end of the decade.

(These are just the most recent in a series of warnings by eminently credible sources dating back to 2004. For some earlier quotes, see the sidebar.)

Now let's take another look at that first study. It says that supply and demand are likely to diverge between 2010 and 2015, *unless demand falls in parallel with supply constrained induced recession.* In other words, this forecast, like the rest, is based on the assumption *that the economy stays healthy,* because (as just happened) an economic downturn reduces the demand for liquid fuels. So we can sum up all four of these recent analyses in one conclusion:

> *IF* the economy stays healthy, *THEN* supply shortages or very high prices will begin to develop before the end of this decade, probably some time between 2012 and 2015.

In forecasting the timing, therefore, the operative question is, *How likely is it that the economy will stay healthy?* And the answer is, *Not very.* This is because fuel prices and the economy have become deeply interdependent. Just as a bad economy causes fuel prices to fall (as we saw in 2008), so high fuel prices cause the economy to fall. An often cited threshold is $85 per barrel, above which the price of fuel has a damaging effect on the economy. Our current economic downturn was about bad credit and a real estate bubble, but some analysts suspect that the first card to be pulled out of the house of cards was the spike in oil prices that briefly drove crude to $147 a barrel.

Instead of the steady decline shown in the EIA graph, therefore, we may see a period of boom-and-bust cycles where a rising economy causes a rise in fuel prices followed by an economic downturn and falling fuel prices. If this happens, the point at which global demand permanently exceeds global supply may, contrary to all the estimates quoted above, be pushed clear into the next decade. *But this does not affect the basic finding that, as a*

society, we will soon use much less liquid fuel, for several reasons.

First, from here on out, both sides of the boom-and-bust cycle limit the amount of fuel we will be consuming on average. Either we will be employed but unable to afford the high fuel prices associated with a good economy, or we will have lower fuel prices in an economic downturn but be unable to buy any because we're unemployed.

Second is the fact that the U.S. imports most of its oil. So for us, the question is not how much oil is being produced globally, but how much of it is available for import. And from this viewpoint, the picture looks very dark indeed. All the big oil exporting countries have internal development needs to meet at the same time that almost all of them are producing less oil every year. The combination of increasing internal consumption and decreasing oil production can very quickly send exports from a given country to zero.

A third factor that guarantees less fuel available to us in the future is China's quiet acquisition of long-term contracts with major oil producers, which will take a lot of oil out of the open market we've been depending on to supply our needs.

Finally, the notion that the global economic cycle will be driven by our national vicissitudes is based on the assumption that the world economy depends on the U.S. economy. That's been true till now, but the moment the Chinese realize that instead of lending us money to buy their products, they can lend themselves the money to buy their products, we fall out of the picture, and at that point we may well find ourselves with a decreasing ability to pay for fuel that is becoming increasingly expensive, with prices driven upward by an Asian economic expansion that has decided to go on without us.

A DANGEROUS SITUATION

The more we consider the dependence of our economy on cheap fuel, the more fragile it appears. Everything about the American economy is based on the assumption that growth is inevitable; indeed, compound interest itself—the bedrock of our financial system—is based on this concept insofar as it represents actual growth and not just inflation. Take that growth away,

and the whole thing collapses, as we saw when real estate prices stopped increasing.

The unprecedented disappearance of spare liquid fuel production capacity makes the system highly vulnerable to interruptions in supply, as diagrammed here by TCLocal contributor Karl North; a problem with oil production (far left) can set in motion a set of feedback loops that brings down the entire economic system. From this perspective, our current situation is actually rather precarious.

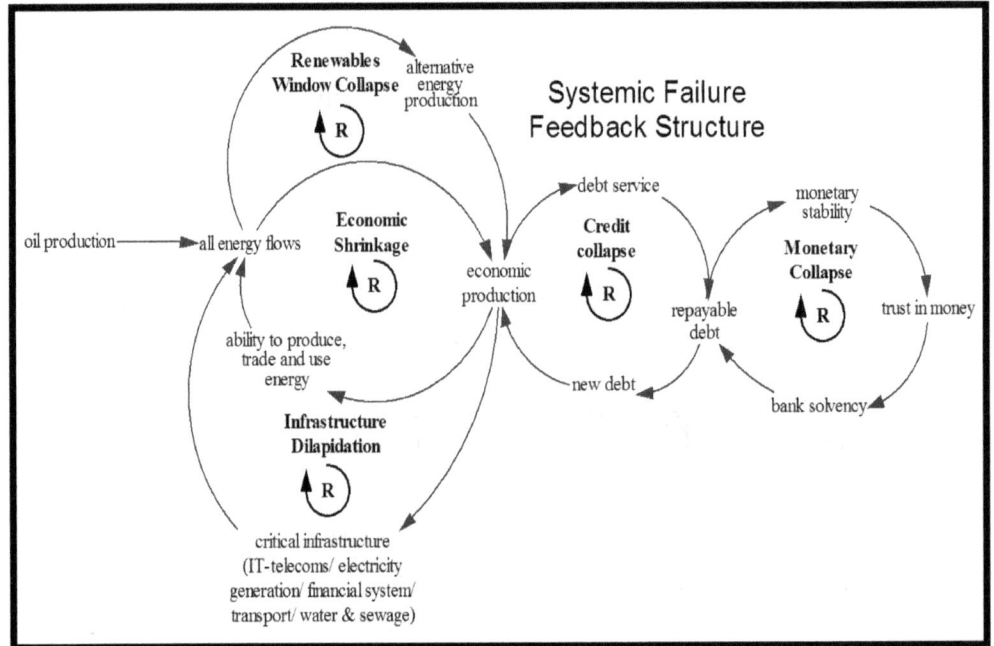

Systemic Failure Feedback Structure

While it's devoutly to be hoped that we can get past the inflection point of oil production with our society more or less intact, no one should underestimate the downside potential of this development. Another recent objective analysis, this one carried out by the German Army (the Bundeswehr), summed up the consequences of declining oil production for their country this way:

> Investment will decline and debt service will be challenged, leading to a crash in financial markets, accompanied by a loss of trust in currencies and a break-up of value and supply chains—because trade is no longer possible. This would in turn lead to the collapse of economies, mass unemployment, government defaults and infrastructure breakdowns, ultimately followed by famines and total system collapse.

There is no reason to believe that the potential damage we could be facing here in the U.S. would be any less than in Germany, which is one of the richest and most advanced countries in the world and one that has put far more effort into transitioning to alternative energy than we have.

THE OUTLOOK FOR THIS DECADE

These considerations lead to the conclusion that the watchword for the coming decade is *instability*. We will probably cycle between economic hardship and high fuel prices for a while, and this cycle will militate against constructive responses. When the economy is bad, we won't have the money to spend on sensible measures like alternative energy and mass transit, and when it starts to recover, we'll tell ourselves that the problem was temporary and that we'll soon be back to business as usual. It's an old story: when the roof leaks, it's raining too hard to fix it, and when it stops raining, a fix isn't needed…until the whole thing comes down on our heads.

As murky as the future appears, however, some things are fairly easy to predict. Here is a list of things that will probably have occurred, or at least be starting to occur, by the end of this decade.

- *Liquid fuels and energy in general will be more expensive.* This one's easy. Even if we could keep expanding oil production (which no one who has looked into it believes), that oil will become increasingly more expensive to extract as we are forced to look farther out into the ocean for it.

- *Less fuel will be available to use.* This is another easy call; either fuel will be too expensive, or we won't be in a position to buy as much as we used to.

- *We will have begun to stay closer to home.* This is already happening. Another way to put it is that life will become more local.

- *Supply chains will have begun to contract.* This is another direct consequence of rising fuel prices. As noted in August 2007 front-page articles in both the *New York Times* and *USA Today*, the distance that goods travel to market became noticeably shorter in just the few months during which we experienced the last price spike. Consequences of supply chain contraction include a shift back to more local production.

- *Food (as a percentage of income) will be increasingly expensive.* Yet another direct consequence of the increasing price of fuel, which is used in enormous quantities both to produce food and to transport it over long distances. Farm land will increase in value, and farm employment will rise as manual labor begins to replace energy provided by liquid fuels.

- *We may begin to see occasional interruptions in some services (electricity, water, sewer, internet, etc.).* This one is not as obvious as the preceding, as none of these services are directly impacted by the price of liquid fuels; but huge quantities of liquid fuel are consumed in *maintaining* all of these service infrastructures, and rising fuel prices will probably result in deferred maintenance and a possible consequent lack of reliability. I don't think this is likely before the end of the decade, but it's certainly a possibility, and one that should be planned for.

- *Rationing of fuel and perhaps even food is possible by the end of the decade.* Rationing would demonstrate real sensitivity for the social justice aspects of the situation, so I don't expect to see it happening any time soon, but it's a possibility.

Some broader developments are simply continuations of current trends that will be accelerated by high fuel prices and their effect on the overall economy.

- *Our standard of living will continue to fall.* U.S. household income in real dollars peaked in 1998-1999 and has been declining ever since. There's no reason to believe that this trend will be reversed.

- *Fewer financial resources will be available to government.* This is another development that's already underway, and it means that most meaningful responses will have to come from individual efforts or self-organized community action.

- *Providing health care for all will be increasingly difficult.* Responses include better health education, free clinics, citizen involvement in county public health advisory boards, and the assumption of greater responsibility for maintaining our own health.

- *Military conflict over resources will become increasingly likely.* Which is, of course, why the U.S. Joint Command is so interested in our energy outlook!

FINAL THOUGHTS

Three observations come out of all this.

The first half of the decade (2010-2015) looks better than the second half (2016-2020). If you have any major projects in mind, this might be a good time to get going. In particular, this would be a good time to make infrastructure improvements, establish a garden, and move closer to work (or arrange to work closer to home).

The developments listed above as possible by 2020 are virtually certain by 2030. The descent doesn't stop until we've achieved a state of equilibrium with a much lower level of resource exploitation. That transition can be easier or harder depending on how we approach it.

A lot of these developments can be prepared for. And that is the purpose of TCLocal: to begin to plan for the future looming on the near horizon. We hope that the foregoing gives the context for our effort and that the articles we've published at tclocal.org are helping us begin to confront and plan for the challenges facing us over the coming decade.

Previous Attempts to Alert Us to the Coming Crisis

The latest predictions of a global oil supply crunch coming somewhere between 2015 and 2020 are in keeping with previous expert findings dating back to 2004 and largely ignored by policy makers. It is unfortunate that it took a massive blowout in the Gulf of Mexico to make people aware that we might have a problem.

The disparity between increasing production and declining reserves can have only one outcome: a practical supply limit will be reached and future supply to meet conventional oil demand will not be available. The question is when peak production will occur and what will be its ramifications. Whether the peak occurs sooner or later is a matter of relative urgency.... In spite of projections for growth of non-OPEC supply, it appears that non-OPEC and non-Former Soviet Union countries have peaked and are currently declining. The production cycle of countries... and the cumulative quantities produced reasonably follow Hubbert's model [which successfully predicted the peak of U.S. lower-48 production more than 10 years ahead of the event].... The Nation must start now to respond to peaking global oil production to offset adverse economic and national security impacts.

Strategic Significance of America's Shale Oil Resource. Vol. 1, Assessment of Strategic Issues. Office of Deputy Assistant Secretary for Petroleum Reserves, Office of Naval Petroleum and Oil Shale Reserves, U.S. Department of Energy, March 2004

[W]hen discoveries of new reserves fall short of demand—which may be the case in just a few years for oil, and slightly later for natural gas—energy prices will climb significantly. The supply situation is being made more acute by the growing hunger for energy in China and India.... The end-of-fossil-hydrocarbons scenario is not therefore a doom-and-gloom picture painted by pessimistic end-of-the-world prophets, but a view of scarcity in the coming years and decades that must be taken seriously. Forward-looking politicians, company chiefs and economists should prepare for this in good time, to effect the necessary transitions as smoothly as possible.

Energy Prospects After the Petroleum Age. Deutsche Bank, December 2004

The peaking of world oil production presents the U.S. and the world with an unprecedented risk management problem. As peaking is approached, liquid fuel prices and price volatility will increase dramatically, and, without timely mitigation, the economic, social, and political costs will be unprecedented. Viable mitigation options exist on both the supply and demand sides, but to have substantial impact, they must be initiated more than a decade in advance of peaking.

Peaking of World Oil Production: Impacts, Mitigation, and Risk Management. Report of Science Applications International Corporation (SAIC) to the U.S. Department of Energy, February 2005

The doubling of oil prices from 2003-2005 is not an anomaly, but a picture of the future. Oil production is approaching its peak; low growth in availability can be expected for the next 5 to 10 years. As worldwide petroleum production peaks, geopolitics and market economics will cause even more significant price increases and security risks. One can only speculate at the outcome from this scenario as world petroleum production declines. The disruption of world oil markets may also affect world natural gas markets since most of the natural gas reserves are collocated with the oil reserves.... The days of inexpensive, convenient, abundant energy sources are quickly drawing to a close.... World oil production is at or near its peak and current world demand exceeds the supply.

Energy Trends and Their Implications for U.S. Army Installations. U.S. Army Corps of Engineers, September 2005

Raising [oil] production is a real challenge... if we stay with this type of production growth our impression is that peak production could be reached around 2020.

Thierry Desmarest, CEO of TOTAL S.A., world's fourth-largest publicly traded oil and gas company. Speech to the World Gas Conference, Amsterdam, 7 June 2006

[W]hile some of the more pessimistic oil specialists are declaring that peak oil has already been passed, or at best is here now, others believe it is not going to arrive before 2010. Some optimists give the world a little more breathing space—that is to say up to 2020, and perhaps even up to 2030. However, all in all, most would appear to agree that peak oil output is not very far away for all of us. It could take place sometime within the next decade or so, which in fact means that there is not much time left for a world economy to be driven largely by oil.... Furthermore, under any of these scenarios, and since peak oil output is not about the time at which oil will run out, but the time at which production can no longer be increased to cope with increased demand, it seems the only way the oil price can go is up.... [W]e are at, or near, the production peak of world oil, if not on the downward slope....

Dr Shokri Ghanem, director of research at OPEC. Speech on the occasion of receiving the 2006 Petroleum Executive of the Year Award, 19 September 2006. Published in OPEC Bulletin

CRUDE OIL: Uncertainty about Future Oil Supply Makes It Important to Develop a Strategy for Addressing a Peak and Decline in Oil Production

Title of report from the U.S. Government Accountability Office to the U.S. Congress, February 2007

I expect to see a peak sometime before 2015, but I don't think we'll see a simple maximum followed by a decline. I foresee a series of maxima, each followed by a brief decline. The simplest analogue would be a sine wave. It may be some time after the true peak before we can recognize it as such.

Jeremy Gilbert, retired Chief Petroleum Engineer for British Petroleum (BP). Interview with ASPO-USA, 7 May 2007

The oil boom is over and will not return. All of us must get used to a different lifestyle.

King Abdullah of Saudi Arabia. Address to his subjects, August 2007

We are experiencing a step-change in the growth rate of energy demand due to population growth and economic development, and Shell estimates that after 2015 supplies of easy-to-access oil and gas will no longer keep up with demand.

Jeroen van der Veer, CEO of Shell Oil. Memo to all Shell employees, 22 January 2008

Can New York State Feed Itself?

By Jon Bosak (June 2009)

For someone who believes, as I do, that decreasing availability of cheap fossil fuel will eventually make the transportation of food over long distances economically unfeasible, the phrase "local food" acquires a special meaning beyond the usual lifestyle implications. It's less about maintaining moral purity and more about whether we're going to have enough to eat. Since I live in the state of New York, the question becomes: could New York feed itself on what it produces?

A couple of years ago, I attempted a back-of-the-envelope sort of calculation to answer this question from a "peak oil" standpoint. To model the worst case, the one in which it takes more energy to extract fossil fuel than the energy we can get out of it, I put the question this way: if New York State produced what it did a hundred years ago, before the arrival of gasoline- and diesel-fueled equipment, could it feed its present population?

The answer, based on New York State agricultural statistics from the 1900 U.S. census, was rather depressing. Despite the fact that New York back then was an agricultural powerhouse—being, for example, far and away the number one state in potato production—its 1900 output of food would barely keep its current population alive.

Carbs weren't so bad; assuming, in round numbers, a state population of 20 million (a little more than the current estimate), NYS 1900 could annually provide each resident with 87 pounds of corn and wheat and 114 pounds of potatoes. But protein was another story. NYS 1900 could provide each current resident with just 16 pounds of beef and pork, 37 eggs, and half a chicken per year. Dairy production, a historical strength in the state, would provide each person now living here just 39 gallons of milk per year, including an average six pounds of butter and seven pounds of cheese. This is probably enough animal protein to sustain life, but not remotely what we're used to.

NYS fruit wouldn't take up much of the slack, either; apples, grapes, peaches, pears, and berries put together would only amount to about 75 pounds per person. New York invented beans as an article of commercial North American agriculture (the first commercial bean crop on record was grown in 1836 in the Town of Yates, in Orleans County), but each person in our current population would only get about four pounds of them a year, plus a little less than a pound of peas. The problem, of course, is that in addition to cutting the fossil fuel input (including all the natural gas we turn into fertilizer), we would be trying to feed almost three times the number of people today that we supported in 1900.

Obviously this calculation was based on some very pessimistic assumptions about available fuel. But it also contained some extremely optimistic assumptions as well—most importantly that we still had substantially more arable land than we actually do now and also that we still had the vastly greater resources of animal power available a hundred years ago.[1] While suggestive, it wasn't a very precise way of assessing our current resources.

THE CORNELL STUDIES

Unknown to me, teams at Cornell University under the direction of postdoctoral researcher Christian Peters were engaged in sophisticated studies that would answer a more immediately interesting question—not what would happen if the energy inputs failed, but what the state's carrying capacity is now, given current rates of production, and what our distribution system would look like if food miles were reduced as far as possible.

The work undertaken so far by Peters *et al.* has been described in two articles published in the journal *Renewable Agriculture and Food Systems*. The first piece, from 2006,[2] investigated the influence of diet on the demand for agricultural land and, secondarily, the ability of New York State to reduce environmental impacts by supplying food locally. The second study, from 2008,[3] focused more closely on local food by developing and applying a method for mapping NYS foodsheds. While

---

1 Anyone who wants to check my figures or apply this method to other states can find a scan of the entire 1900 census abstract at http://www.ibiblio.org/tcrp/src/1900census.pdf (this 66 MB file is best downloaded before viewing).

2 C. J. Peters, J. L. Wilkins, and G. W. Fick, "Testing a Complete-diet Model for Estimating the Land Resource Requirements of Food Consumption and Agricultural Carrying Capacity: The New York State Example," *Renewable Agriculture and Food Systems* 22(2), pp. 145-153.

3 C. J. Peters, N. L. Bills, A. J. Lembo, J. L. Wilkins, and G. W. Fick, "Mapping Potential Foodsheds in New York State: A Spatial Model for Evaluating the Capacity to Localize Food Production," *Renewable Agriculture and Food Systems* 24(1), pp. 72-84.

preliminary, the results of these studies pose serious questions for those who seek to relocalize our diet, and they raise some significant issues for planners attempting to grapple with the contraction of agricultural supply chains due to rising fuel prices. The purpose of this article is to make the key findings of these seminal studies available to a larger audience.

The relatively short list of products actually produced in our climate suggests that the answer to the question of how much of our food needs can be supplied locally depends to some extent on what kinds of foods we plan to eat. The 2006 study approaches this issue by using USDA data to define 42 different nutritionally complete diets supplying 2300 calories a day, calculating the agricultural land requirements for each diet, and then calculating the potential ability of NYS to supply that diet to each resident based on recent estimates of available agricultural land (not land currently in production, but land that could be). Each of the 42 diets is nutritionally complete but contains different proportions of meat and eggs at rates from 0 to 12 ounces per day and different proportions of calories from fat ranging from 20 to 45 percent of total calories. The average U.S. diet contains 5.8 ounces of meat or eggs per day and 41 percent of calories from fat; Figure 1 shows where this average diet falls in the six by seven matrix formed by the two variables.[4] While obviously incomplete, the model does represent the range of common American food consumption patterns from low-fat lacto-vegetarian to high-fat, meat-rich omnivorous.

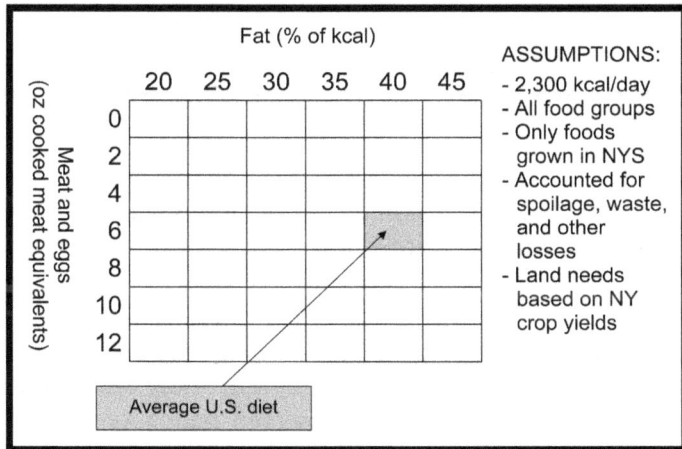

Figure 1. Matrix of 42 complete diets

Land requirements for each diet are based on a division of available agricultural land into three categories: harvested cropland, cropland pasture, and permanent pasture.

4 Except for two screen shots (Figures 7 and 8), all the illustrations in this article come from a presentation given by Dr. Peters at the conference "Planning for Farms, Food, and Energy in Central New York" sponsored by the American Farmland Trust 25 March 2009 in Syracuse. I am indebted to conference organizer Judy Wright for a copy of the presentation slides.

LAND USE	AREA 1999-2003
	10^6 hectares
Land in farms	3.12
Total cropland	2.00
Harvested cropland	**1.52**
Cropland pasture	**0.26**
Other cropland	0.23
Permanent pasture	**0.26**
Woodland	0.62
Other land	0.24
Land available to model	**2.03**

Figure 2. Available agricultural land in New York State

Further methodology, detailed in the study, addresses the interdependencies between perennial crops (grown mainly on grassland) and annual crops (grown mainly on cultivated land), and the calculation of carrying capacity employs a conditional equation that determines which category of land is limiting to food production. Figure 3 shows the results, with the seven levels of meat consumption displayed across the bottom and the six levels of fat consumption grouped within each meat consumption level. For example, someone who ate 190 grams (6.7 ounces) of cooked meat equivalents per day would require somewhere in the neighborhood of 0.45 hectares (about 1⅛ acres) of combined annual and perennial NYS crops for their sustenance if their entire diet came from within the state.

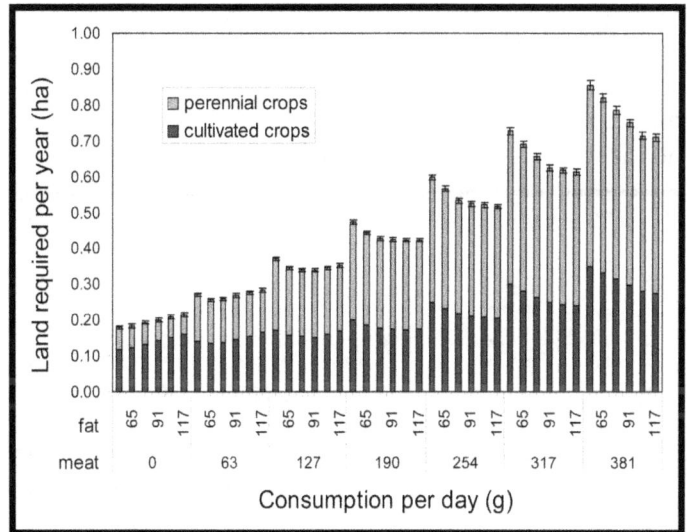

Figure 3. Land requirements of complete diets

EFFECT OF DIET ON CARRYING CAPACITY

Not surprisingly, the results show a nearly fivefold difference in the amount of land needed per capita depending on the diet, from 0.18 ha (0.44 ac) for a diet of 0 g meat and 52 g fat to 0.86 ha (2.12 ac) for a diet of 381 g meat and 52 g fat. As most TCLocal.org readers are aware, animal products require much more land per unit of edible energy than grains; in NYS this amounts to 3.3 to 6.3 times as much total land required for the animal products other than beef and a whopping 31 times as much for beef.

On the other hand, as shown in the figure, much of the difference is in the amount of land devoted to perennial crops rather than cultivated crops. If we consider just cultivated land requirements, the clear animal products winner is whole milk (1.2 square meters of cultivated land per 1000 calories). This is just slightly above the figure for grains (1.1 square meters per 1000 calories) and actually below the requirements per 1000 calories for oils (3.2 square meters), pulses (2.2 square meters) and even vegetables (1.7 square meters).

Beef is always presented as the bad boy in discussions of agricultural requirements, but this seems to depend on where you are. The fact is that a lot of the NYS agricultural land base is not suitable for the production of annual crops but is great for forage, which provides most of a ruminant's nutritional needs. Grassland (I will note) also requires much less in the way of fertilizer and energy inputs and helps to conserve topsoil and nitrogen. Most other foods, including most other animal products, require annual crops, the land for which is more limited in extent and is therefore the limiting factor in the total NYS food supply. Using NYS production figures, the study finds that beef (all cuts) requires 5.3 square meters of cultivated land per 1000 calories, whereas pork (all cuts) requires 7.3 square meters and chicken (all cuts) 9.0. The energy implications of these findings are not brought to the fore in the articles under review here, but clearly the effect on total production and energy requirements of including various kinds of meat in the diet is to some extent location-specific and not as straightforward as it's often assumed to be.

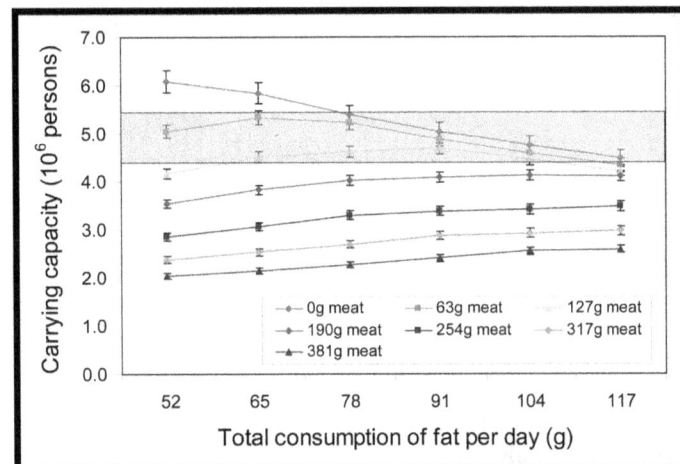

Figure 4. NYS carrying capacity according to diet

Another nonobvious outcome that can be seen by studying the different fat proportions for each meat consumption level in Figure 3 is that increasing the amount of fat in the diet somewhat reduces the amount of land required. As a result, the difference in carrying capacity due to differences in diet is closer to threefold rather

than the fivefold difference suggested by Figure 3. This is summed up in Figure 4, which shows the potential carrying capacity of the NYS agricultural land base for each of the 42 diets. In general, the population supported by NYS decreases with increasing fat in the no meat diet, reaches a peak and then declines in the 63 and 127 g meat diets, and increases with increasing fat in the 190-381 g meat diets. As indicated by the grey shading, some diets with low to modest levels of meat feed equal or greater numbers of people than lacto-vegetarian diets with moderately high levels of fat.

One possibly unexpected implication of the study is that a vegan diet does not support the maximum number of people, at least not in the state of New York: "[W]e conclude that the inclusion of beef and milk in the diet can increase the number of people fed from the land base relative to a vegan diet, up to the point that land limited to pasture and perennial forages has been fully utilized." Figure 5 shows what's meant by this; even the diet with the highest proportion of meat still doesn't exhaust the land available for forage.

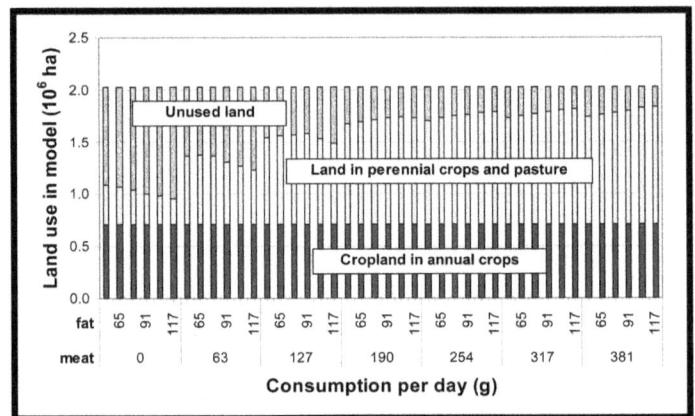

Figure 5. Use of available NYS agricultural land by diet

In a passage sure to provoke some of our readers, the authors continue: "[T]he higher populations supported by lower fat, non-vegetarian diets relative to higher fat, [lacto-]vegetarian diets support the claims by animal scientists that the inclusion of animal products in the diet can increase the amount of humanly edible calories available in the food supply. Indeed, more substantial differences may have been observed had a vegan diet been included among the diet scenarios." The authors hasten to add that this is not an endorsement of the average American diet: "Nonetheless, it is critical to note that the area of overlap observed occurs between 63 g (2 oz) and 127 g (4 oz) of meat, far below the 163 g daily consumption of the average American."

Beyond these details, Figure 4 also provides the answer to my original question: Can NYS feed itself? The answer is an unequivocal No. Assuming that everyone gets a complete, balanced daily diet that includes 190 g of meat and contains 30 percent fat, the state could potentially feed about 21 percent of its current population. Given a radical change in the average diet, this proportion could, judging from Figure 4, rise to a little over 30 percent, but it's clear that NYS will always be a net importer of food. Since the cost of transporting food from outside the state is certain to increase dramatically over the next couple of decades, the effect on food prices can readily be imagined. I think this also suggests that economic forces will push back into production some land no longer considered agricultural (golf courses, lawns, etc.).

A subsidiary but still interesting question for people living out here in Tompkins County is whether the situation just described is the same for all parts of the state; after all, a basic (if mostly tacit) assumption of relocalization is that things aren't going to be the same everywhere. Peters *et al.* address this question in the second of the two articles reviewed here.

FOODSHEDS

The 2008 paper takes on the question of what we mean by "local" in an increasingly urban civilization. "To what degree *can* food be produced locally?," the study asks. "Moreover, should the meaning of 'local' be context specific?" The method is based on a relatively recent reintroduction of the concept of a foodshed, first used by W.P. Hedden in 1929. Peters *et al.* define a *potential local foodshed* as "the land that

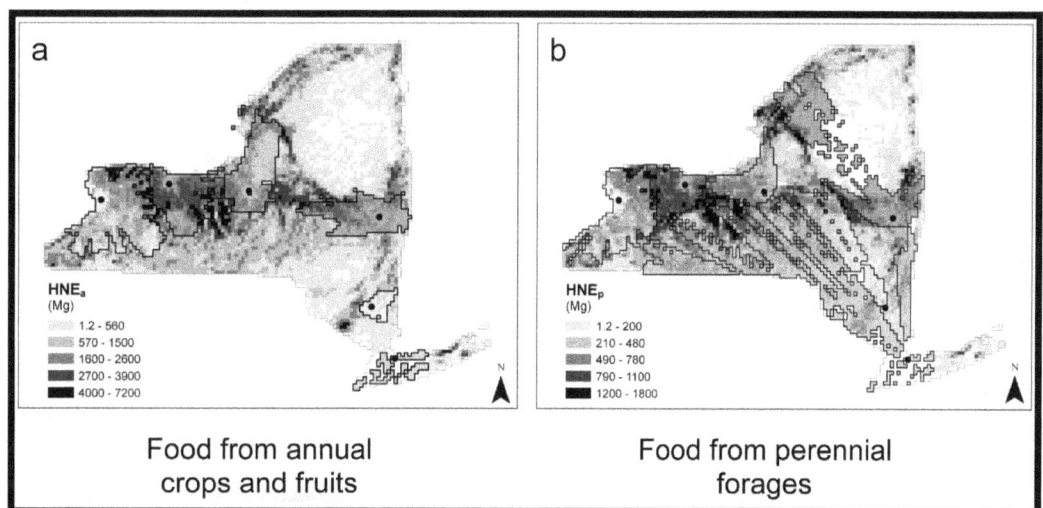

Figure 6. Statewide maps of selected foodsheds (see the original article on tclocal.org for color versions)

could provide some [specified] portion of a population center's food needs within the bounds of a relatively circumscribed geographic area," or more simply, "the area of land that feeds, or could potentially feed, a population." Foodsheds provide a framework for analyzing the capacity to produce food locally at the scale of an individual city, and a principal goal of the 2008 study is to develop standard methods for this kind of analysis.

The model created in support of this goal employs geographic information systems (GIS) to estimate the spatial distribution of food production capacity relative to the food needs of a given population center and then applies optimization tools "to allocate production potential to meet food needs in the minimum distance possible." The software implementing the model also produces foodshed maps that aid in visualizing the geographic extent of a food supply.

Assuming a constant basis in the land use data from NYS, it's apparent that studies of this kind will produce different results depending on the assumptions regarding nutritional requirements and the algorithms built into the foodshed optimization technique.

Since the focus in the second study is on foodsheds rather than dietary variables, it holds those variables constant by using just a single representative complete diet containing 6 ounces daily from meat and eggs and 30 percent of calories from fat. A number of other simplifying assumptions are needed to make it possible to do the spatial modeling; for example, because the concept of a foodshed is tied to population centers, rural NYS residents are assumed to get their food from the nearest center. Also, and crucially, the model seeks to find the minimum total distance food would optimally travel throughout the state rather than optimizing for an individual population center, since the most efficient al-

location for the whole state might require that land near one population center be assigned to a more distant population center. Due to matrix size constraints imposed by the spreadsheet software, only 125 of the 132 statistical NYS population centers could be included in the model, resulting in the elimination of the seven smallest (totaling just 0.2 percent of the state's population).

Even with these simplifications, the optimization model used to calculate foodsheds is quite complex, and I'll have to refer readers who want more details to the published study itself.

A selection of the output produced from the model for the largest NYS population centers is shown in Figure 6.

These maps show foodsheds for food from annual crops and fruits (on the left) and food from perennial forages (on the right) for the six largest consumption zones in New York State: Buffalo, Rochester, Syracuse, Albany, Poughkeepsie-Newburgh, and NYC. These six foodsheds, indicated by different colors (see the original version of this article at tclocal.org), are layered over greyscale shadings showing the capability of different areas of the state to produce food. For example, the completely black pixels in the map on the right show that the area represented by those pixels in the original model (not necessarily scaled the same as the pixels here) is potentially capable of producing 1200 to 1800 metric tons (Mg) of food products annually from perennial crops, chiefly pasture. HNE stands for "human nutritional equivalent," referring to a complex submethodology for relating per capita nutritional requirements to combinations of farm products.

As can be seen from these maps, the presence of a population center much larger than the rest changes the shape of the other foodsheds. For example, on the perennial forages map, the Syracuse, Albany, and Poughkeepsie-Newburgh foodsheds extend farther to the north and west than to the south and east because the *overall* statewide food travel distance is shortened by ceding the land to the south of these centers to the NYC foodshed. This distortion takes an extreme form in the case of the Poughkeepsie-Newburgh foodshed (yellow in the online version of this illustration), which extends from the population center as if it were being blown back by the enormous NYC food demand. Conversely, when a population center is relatively isolated, as in the case of Rochester and Buffalo, its potential foodshed spreads more evenly because it is limited by natural barriers rather than by competition with other cities.

This single example doesn't begin to do justice to the resource provided by the model. I urge people interested in exploring the model further to check it out online:

http://www.cals.cornell.edu/cals
/css/extension
/foodshed-mapping.cfm

OUR LOCAL FOODSHEDS

Below are screen captures of two maps generated by the Cornell tool for the Ithaca foodshed, one map for cropland (annual crops) and one for grassland (perennial crops), with the outlines of the corresponding Syracuse, Binghamton, and Elmira foodsheds shown for comparison.

According to the model, in a distribution system that used all available NYS agricultural land, provided a certain balanced diet to everyone, and optimized statewide food distances, Ithaca's food from cropland (Figure 7)

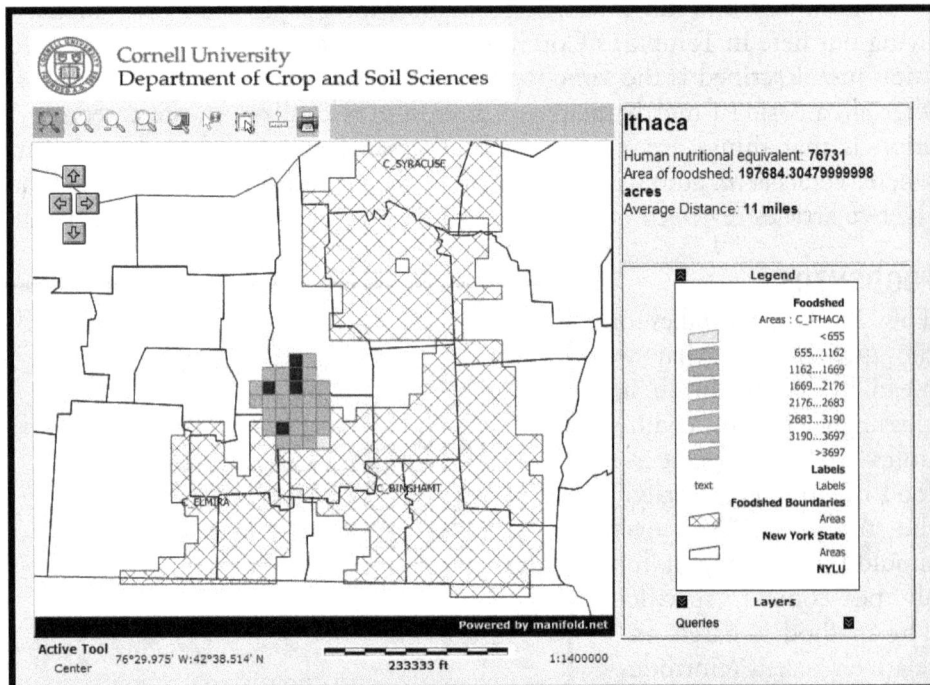

Figure 7. Potential optimized Ithaca cropland foodshed

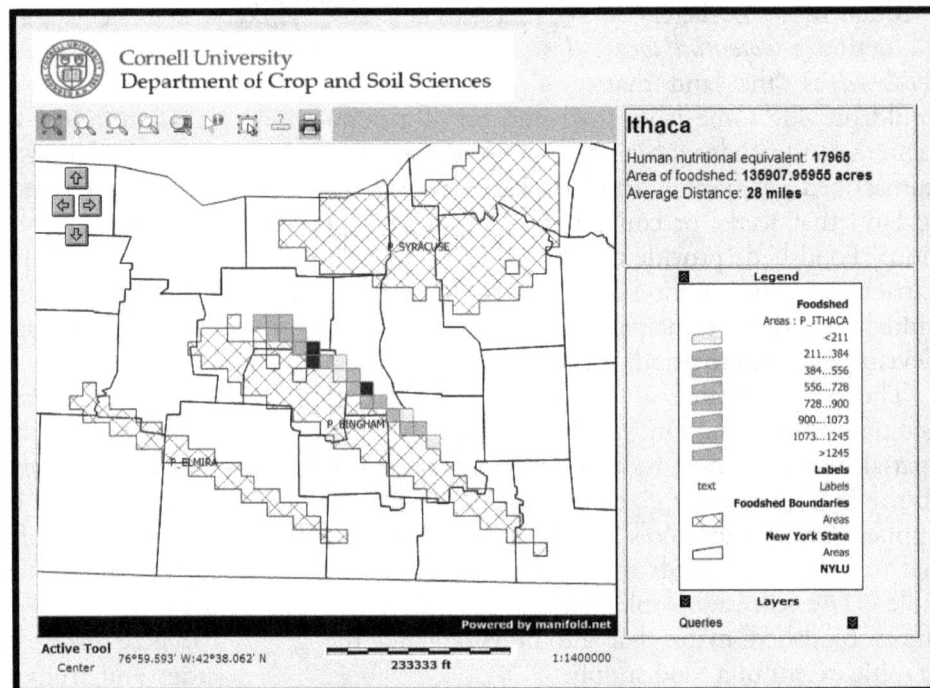

Figure 8. Potential optimized Ithaca grassland foodshed

14

would travel an average of just 11 miles, and its food from grassland (Figure 8) would travel an average of 25. In neither case, however, would that locally sourced food satisfy all the food needs of the Ithaca area population (estimated at 95,000 persons, which includes the Ithaca Urbanized Area plus nearby surrounding rural populations). The model shows that the optimized locally sourced food from cropland would fully supply the cropland component of the assumed diet for about 81 percent of the local population (76731/95000), whereas the locally sourced food from grassland would supply only about 19 percent of that dietary component (17965/95000). This illustrates in detail the conclusion reached in the TCLocal.org article that Dr. Peters published here in April: only about half of our food supply in Tompkins County would come from local sources if food was distributed in a way that minimized food miles for the entire state.

The effect of the immense NYC demand for food on the shape of our optimized foodshed is clear even at this distance from the city; both of Ithaca's foodsheds lie entirely to the west and north of the population center, extending in the case of grassland across several adjacent counties. Also apparent from these maps is the basis of the model on *potential* agricultural land rather than the land that's in production right now; anyone familiar with the areas included in these foodsheds knows that in fact much of the land shown as the potential source of our local food is not now actually in production. The need to preserve currently idle agricultural land north and west of Ithaca for future use has important implications for zoning and land use policy in our area; as the cost of transportation grows, this is where much of our food will have to come from.

WHERE TO BE A LOCAVORE

The table below provides one answer to the question, "how much of New York's food can be provided locally?" The answer is: it depends on where you live.

Population center(s)	Food allocated HNE$_t$		Food distance HNE$_t$	
	Tg	*% of need*	*Tg-km*	*km*
NYC	0.33	2.2	88	264
Urbanized areas	4.58	83.7	233	51
Urban clusters	3.22	98.4	81	25
TOTALS	*8.13*	*34.4*	*402*	*49*

Figure 9. Summary of model output

The table lists three categories of NYS population centers (using terminology from the U.S. Census) in order of the amount of food in Tg (millions of metric tons) they receive within the model. First, of course, is New York City, which is in a category by itself. In this model

—which, it must be remembered, optimizes food distances for the whole state—NYC would get just 2.2 percent of its total from food produced within the state, and that food would have to come from an average of 264 km away. The next biggest population centers, the "urbanized areas," would get 84 percent of their food from inside the state, and that would come on average from 51 km away. And the smallest population centers (excluding the seven very smallest, as noted above), could get virtually all their food needs met from within the state, and the food could come on average from just 25 km away.

Bottom line for the state as a whole: Given the diet assumed for the study, if all agricultural land were in use, and food distribution were optimized to minimize the total distance that food travels, New York State could get 34 percent of its food needs met from within the state, and that food would travel an average distance of 49 km to each consumer.

You'll notice that the 34 percent figure differs a little from the results of the 2006 study, due no doubt to differences between the two studies in assumptions and methodology. The difference isn't enough to change the basic picture and in fact reinforces it by coming at it from a different angle, but it's obvious that the results provided by a model like this depend to a large extent on a complex set of assumptions. The authors point out several ways in which the model does not take into account real-world factors (geographic limitations, agricultural specialization, details of the food processing workflow, economies of scale, etc.) and note that optimizing for food miles does not necessarily optimize for greenhouse gas emissions or energy inputs. Nevertheless, one conclusion stands out fairly clearly. Outside of the NYC area, most population centers in the state could meet all, or nearly all, of their needs from food produced within the state. But NYC, if it depended on food produced within the state, would go largely unfed.

The study boils the results down to what I would call the good news and the bad news. The good news is that "NYS may be able to significantly reduce the distance food travels" to an average far less than the 1300 miles often cited as the distance from farm to consumer in the U.S. The bad news is that "feeding big cities may require food to travel great distances."

Peters *et al.* don't draw out the implications of that last point, but I will: People living in NYC are going to be paying an awful lot more for food as we begin to move down the energy descent slope, and it would be better for them if they started to relocate back to the small towns upstate that have seen their populations decline over the last half century. To rephrase the old saying, NYC is a nice place to visit, but I wouldn't want to try to survive there.

E

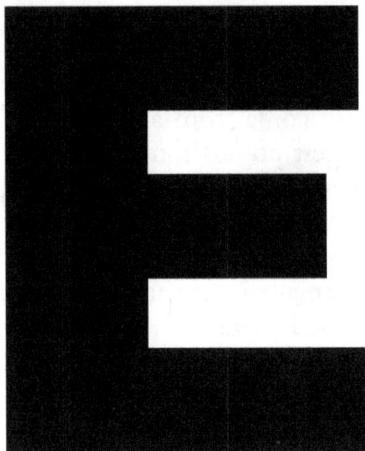

Envisioning Tompkins County Food Production

By Karl North

Editor's Note: This article was originally a six-part series published at intervals from July 2009 to July 2010.

INTRODUCTION

In this paper I will attempt a preliminary vision of a re-localization of food production designed to feed the population of Tompkins County. A project of this scope implies a reorganization of food processing and distribution that, while not included in this first iteration, will need to be integrated in a later, expanded overview.

My purpose is to explore the kind of local food system that will be needed as this country faces sharply lower access to the energy sources on which our present industrial form of agriculture and food economy heavily depends. I will describe the types of local farming enterprises, farming methods, resources, and land use needed to confront a future of much lower energy use. A documented baseline assessment of current food production and county resources is not an objective of this essay, but will be essential to a detailed planning effort. The picture presented here is intended to be general enough to be useful in planning the relocalization of foodsheds that include an urban center the size of Ithaca, New York.

In these first few pages, I will set out my premises and theoretical points of departure in some detail to explain the fundamental changes in perspective I think are necessary to envision how and where we produce food in the future.

This vision will rely on several critical premises:

1. The premise underlying all work of TCLocal is that a permanent decline in the availability and affordability of liquid fuels and related rising costs of all energy sources will inevitably lead to much lower energy use and increasing importance of local scale in human af-fairs. The present long-distance food economy will shrink, and consumers will need to rely increasingly on local food production.

2. This "energy descent" will force the transformation of food production toward low external input systems that rely more on human labor and models of healthy, highly productive ecosystem processes common in nature instead of the high energy cost technological substitutes on which agriculture, including most of organic agriculture, depends today.

3. Our world is systemic in nature (parts are more or less connected), and this has important implications for attempts to change it. Problems we want to solve are, as the system analysts like to say, "structural," and require intervention in several places. So the single-issue approach to any kind of change is eventually bound to fail to meet expectations. For example, dieting to solve weight problems never works for long if the problem lies in the structure of our life. In addition to changing what we eat, maybe trading the car in on a bike and some tools to dig the lawn into a vegetable garden would produce better results. By itself, widening Ithaca's commuter feeder roads like Route 13 will not solve the traffic problem; the improved highway only attracts more cars. But it might succeed if coupled with a county tax on car ownership, a tax hefty enough to pay for major improvements in public transportation. This would be intervention in the very "structure" of the county transportation system.

Moreover, despite best intentions, in a systemic world we can never make just the one change we aim for. Complex systems are squishy like a balloon: squeezing one end only makes the balloon blow out in other unexpected places. Change agents need a holistic approach that recognizes that consequences of any interventions are multiple ripple effects distant in space and time. This approach has important implications for design at every level of scale.

In complex systems, cause and effect are often distant in time and space

Figure 1. Unintended consequences

At the garden or farm scale we want to build in *multifunctionality,* where parts of the system serve more than one purpose. Plants and animals that provide food, for example, may also provide ecological services necessary for the health and productivity of the whole. Ecological services are the benefits arising from the functioning of the ecosystem, in contrast to purchased inputs.

At the level of the *food system,* where different elements of production, processing, and distribution can be designed as a cooperating whole, we need to build in *complementarity* as to what is produced, and services that are shared among the different types of production units to be described in this paper. Urban gardens may best serve the county food system by growing fresh produce, thus complementing rural farms that produce less perishable foods, for example.

At the community level, we need to view the reorganization of the food system as affecting and affected by the reorganization of all other infrastructure and institutions impacted by reduced energy availability, e.g., industry, housing, markets, transportation, sanitation, information flow, knowledge production, etc.

Most important from a systems perspective, we need to regard far-reaching changes like those to be proposed here as experimental, and track for unintended consequences in time and space. This approach, known to ecologists and other systems thinkers as *adaptive management,* requires constant monitoring and replanning in the face of uncertainty about consequences.

4. The design of a relocalized agricultural system will need to address root causes. For example, the *proximate causes* of flooding may be failed riparian buffers and levees, but the *root causes* are pavement, bare ground, and other surfaces that create surface run-off, soils compacted and depleted of water-holding organic matter, agricultural field drains, and channeling that cuts streams and rivers off from their historic flood plains. Attention to root causes forces the need for the systems perspective outlined in premise #3. If, from the viewpoint of sustainability, high-input, oil-dependent agriculture is now revealed to be a design failure from the outset, little is gained by piecemeal solutions like replacing chemical inputs with "natural" ones. Rather than the input substitution approach, efforts are better directed toward whole agroecosystem design that integrates a diversity of spatial and temporal elements.

Understanding Sustainability. In addition to working from the stated premises, I want to ground the proposals in this visionary project in a working concept of sustainability based on ecological science. This is important at this historical juncture for a couple of reasons. The common practice of confusing and conflating sustainable agriculture and organic agriculture will be counterproductive in the coming era when shrinking access to cheap energy will reveal the unsustainability of most current forms of agriculture, including organic. The flowering of the organic farming movement, in which I have been a practitioner for 30 years, generated much innovation that will be useful in coming years. But it also produced the delusion of a luxury version of sustainability, because it occurred in and was shaped by an era of cheap oil. Limited by economic forces and a focus mainly on environmental issues, organic farming became more a matter of substituting "greener" inputs for those of industrial agriculture rather than seeking input independence through systematic redesign. Awareness that many of the "greener" inputs depend on fast-depleting, often finite, soon-to-become-expensive resources still has not penetrated the organic movement sufficiently to become a paramount concern. A common practice in organic vegetable farming, for example, is to import fertility in the form of compost from factory-style dairy and poultry farms.

None of the above should be construed as an attack on the organic farming movement, or a dismissal of its contributions to the development of a truly sustainable agriculture. But we need a more rigorous design tool than "organic" to select from those offerings.

Sustainability means that local food production systems must support the food and fiber needs of a given human population without exceeding their carrying capacity (CC). A working definition of CC might be *the maximum indefinitely supportable ecological load of an ecosystem or area.*

We must be clear about what constitutes a *supportable ecological load.* Depletion of a finite resource like copper or phosphorus is not supportable unless we find a way to perfectly recycle as much of it as is needed (not downcycle it as in plastic bags → park benches → landfill). Petroleum products used for fuel are not recyclable, and anything needing those fuels in its production is therefore unsustainable. The supportable load on renewable resources on which we depend is limited to their refresh rate. The rate at which a farm consumes soil organic matter depends on the capacity of the agroecosystem to rebuild it. Less evident, but perhaps ultimately most important, is the load of work we place on natural systems to absorb concentrations of substances and handle imbalances that we create. That load can become insupportable, either because it becomes too great or because we weaken the ability of natural systems to do the work. In short, the success and survival of all human activity rests on and must be subordinate to the continuing health of the natural resource base and the ecosystems that underpin it. Encapsulated in the phrase, "Mother nature bats last," this means that any sacrifice of ecological health to advance human affairs eventually results in losses to society. Economic profit gained in the short term at the expense of the natural resource base and its health leads inevitably to economic loss in the long term.

The CC of a specific farm or regional landscape at a given historical moment may have eroded far below its potential. Industrial agriculture has indeed damaged the CC of much of the agricultural resource base. At present, technological props based on cheap oil have created a temporary, artificially higher CC that ecologist William Catton called "phantom carrying capacity."[1] Continued belief in this phantom can prolong the overshoot and erosion of real CC long enough to cause the population to collapse. Our present food system is operating at phantom CC. This is due to a level of agricultural productivity that is temporarily and artificially high because it relies on fossil fuels and other raw materials that are finite and fast depleting. Over 80 per cent of the energy on which our food system runs comes from oil. In practical terms this means that we are feeding more people than is sustainable (at least on a global basis), because human populations have ballooned in response to rising food production. Equitable food distribution is an essential response to the problem but is ultimately insufficient unless agriculture itself can be organized on a sustainable basis.

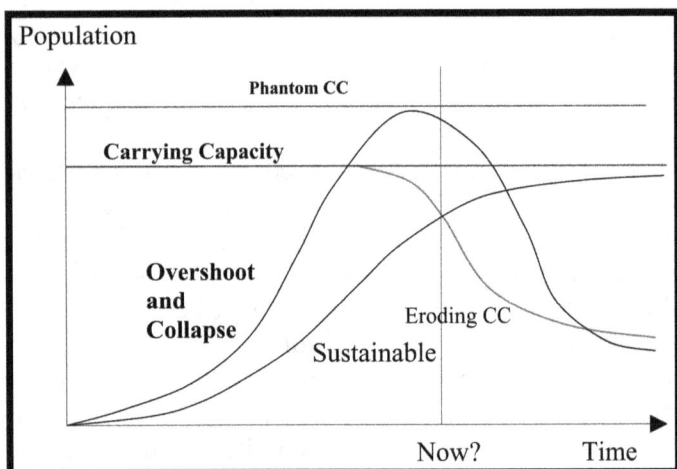

Figure 2. Overshoot

On the other hand, human intervention can often rebuild CC and possibly improve it somewhat. Effective agroecosystem design can improve farm sustainability, for example, by building in sufficient species diversity to provide necessary farm inputs and ecological services "for free" to replace unsustainable external inputs to farms.

Finally, "needs of a given human population" is a slippery term, the definition of which varies widely from one culture to another. We need to ask: How much material consumption does our quality of life really require? In regard to food, does discretionary consumption exist which, if reduced, could allow agriculture to feed more people?

Despite the complexity of these questions, thinking about sustainable design to respect carrying capacity has effectively focused the attention of ecological scientists on maximizing the long-term health of four interrelated ecosystem processes in agroecosystems:
1. The mineral or nutrient cycle
2. The water cycle
3. The energy flow
4. The structure and interactions of the biological community

A focus on these four processes leads to the development of principles or attributes of sustainable agroecosystem design intended to maintain, or in many cases regenerate, the health of these ecosystem processes. Some of the widely accepted principles and their implications are:

- **Low external inputs** — Input self-sufficiency.
- **Low emissions** — Closed nutrient and carbon cycles that avoid losses of valuable resources that eventually cause environmental damage.
- **Stability – Resilience – Adaptive Capacity** — These qualities of sustainability are all necessary, but since they exist somewhat in tension, there must be balance among them. Stability is the quality that produces reliable results and minimizes risk, but in excess, stability can become rigidity. However, a certain flexibility is required for resilience, which is the ability to rebound from sudden change like a dry period in the farming season. Adaptive capacity to respond to slower changes like a gradually invasive plant disease also requires flexibility. Reserves of material or energy, overlaps, redundancy, or other slack in a system provide that flexibility, but at the price of efficient use of resources.
- **Knowledge intensity** — Reliance on technologies that are powerful but derivative of a narrow, specialist knowledge base will give way to a broader, more demanding knowledge of farms as complex ecosystems of interdependent species, a knowledge that enables the creation of biodiversity to capture synergies, to biologically control pests, for example.
- **Management intensity** — Farming for input self-sufficiency and low emissions will require more labor devoted to management planning and monitoring to replace other resources or use them more efficiently to maximize sustainable yield: productivity per acre.
- **Local food self-sufficiency and national food sovereignty**

---

1 William R. Catton Jr., *Overshoot: The Ecological Basis of Revolutionary Change* (Urbana and Chicago: University of Illinois Press, 1982).

These principles fit well with the design imperatives of a future marked by gradual loss of sources of cheap energy. Aimed at maximizing the ecosystem processes described before, these design principles will guide the visioning effort.

The visioning process will draw on several main resource areas:

- Known principles of agroecology and their relation to the concept of sustainability as outlined above;

- Historical knowledge of how production was achieved before the era of cheap energy and other inputs—as late as the early 20th century in some locations;

- Subsistence and semi-subsistence farming systems in agrarian communities on the periphery of the global industrial economy, which have managed to escape the imprint of the current system;[2]

- Contemporary models of large-scale conversion from industrial agricultural systems to localized, low input agricultural systems as in Cuba,[3] the resources of the Permaculture[4] and Transition Towns[5] movements, and some of the more sustainable design efforts to develop very low external input systems in the organic agriculture movement.

From these resources I will attempt to extract and introduce a set of general food production system design strategies that follow principles already outlined above. Many of their elements have in common the goal of designing for food and other species that are multifunctional, delivering ecological services presently provided by the external inputs in our industrialized food system that will become prohibitively expensive in the future. Elements of these food system design strategies include:

1. Integration of crops and livestock

2. Animal, human- and small-scale wind, hydro, and solar as the primary energy sources of agricultural production

3. Perennial crop polycultures, in particular, perennial carbohydrate crops(nutritionally, hazelnuts can be seen as equivalent to soy, chestnuts as an equivalent to corn)

4. Perennial forage polycultures under intensive management (variations on an interdependent triad: grasses for bulk, legumes for nitrogen, deep-rooted broad-leaf forbs for minerals)

5. Agroforestry and sylvopastoralism

   a. Alley cropping/grazing within perennial polycultures

   b. Terracing, or return of perennials to erodible slopes

6. Intensive water management: capture and distribution swales, rooftop capture, microclimate creation, ponds and filter wetlands for storage, nutrient processing and aqua-ecosystem development

7. Extended growing season and harvest technologies

8. Intensive nutrient management

   a. Repairing and tightening broken and leaky nutrient cycles: food = waste = food

   b. Rotations that manage nutrient capture and use

9. Intensive bed growing

10. Biocontrol of pests: pest predator production and habitats, trap crops

11. Plant families designed for symbiosis

12. Stacked species for sunlight capture or shade or wind protection: vertical plant growth—vine crop fences, espalier

13. Cooperative management: neighborhood and community gardens, revival of the commons

Historical models of energy-efficient foodsheds that include an urban population suggest the need to design a whole that integrates three somewhat overlapping categories of production systems:

- **Urban agriculture** — many small islands of gardening in the dense city center

- **Peri-urban agriculture** — larger production areas in the immediate periphery

- **Rural agriculture** — feeder farms associated with village-size population clusters in the hinterland of the city but close enough to be satellite hamlets

The design of each type of system will vary depending on its available resources, its appropriate role in feeding the county population, and its input support function for the other production categories.

GENERAL PROBLEM AREAS IN SUSTAINABLE AGRICULTURAL DESIGN

Four key issues must be addressed in order to envision these three county food production systems: fertility, energy, water, and pest control. But first, a word about the role of species diversity in addressing these issues.

In an energy descent environment, agriculture that incorporates the necessary diversity of species that are multifunctional—providing both ecological and other

2 Veronika Bennholdt-Thomsen and Maria Mies, *The Subsistence Perspective: Beyond the Globalized Economy* (London: Zed Books, 1999).

3 Fernando Funes et al., *Sustainable Agriculture and Resistance: Transforming Food Production in Cuba* (Oakland: Food First Books, 2002).

4 Bill Mollison, *Permaculture: A Designer's Handbook* (Tyalgum, Australia: Tagari Publications, 1997). Examples: http://www.youtube.com/watch?v=Bw7mQZHfFVE&NR=1

5 Rob Hopkins, *The Transition Handbook* (White River, Vermont: Chelsea Green Publishing, 2008).

services and food—will gradually replace the current agriculture that substitutes external inputs to solve these problems.

Some of the most durable and productive low input farming systems in history are designed around animals that can accelerate the growth and conversion of plants to fertilizer. Because they are highly multifunctional, ruminant mammals rank highest among these. Beyond their manure production function, they can consume fibrous perennials unusable for human food. These perennials can grow on hill land too rocky or too erodable for food cropping. Used as work animals, ruminants multiply the energy input from human labor many times. They provide a source of concentrated protein food that can be conserved and stockpiled for winter consumption. They provide hides and fiber for clothing as well. Cattle, sheep, goats, alpacas, llamas and bison are ruminants that we can most easily use in agricultural systems in our environment.

A few other animals serve some of these functions and, properly integrated, often are found enhancing these systems. Pigs and poultry can do the hard labor of turning manure into compost, and can thrive by consuming unused and pest species as well as waste streams from farms and kitchens. They both can reduce a patch of weeds to bare ground ready for planting, and pigs will perform tillage as well. They will consume crop residues and garbage from food preparation, and convert it to fertilizer as well as their own production as food animals. Poultry will consume weeds and insect pests. Edible fish and other water animals like frogs and snails can perform the same functions in aquatic systems. This map of flows among components demonstrates the potential of integrated systems (Figure 3). Notice that the flows may go in both directions among all components:

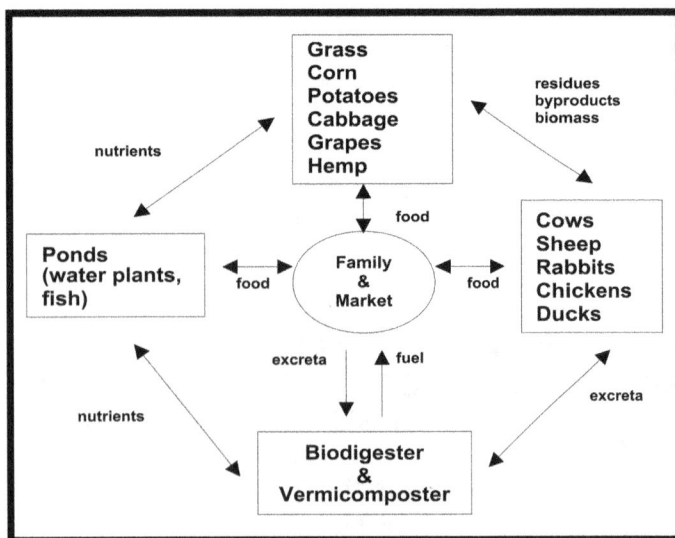

Figure 3. Dynamics of a hypothetical sustainable system

1. Soil Fertility

As energy descent deepens, two key fertility crutches of industrial agriculture will become cost-prohibitive. Synthetic nitrogen fertilizer production requires large quantities of energy. The decreasing quality of phosphate deposits is already driving up the price of phosphate fertilizer (up 700 percent in a recent 14-month period) and production is estimated to peak within 20 years.[6] Moreover, the affordability of most off-farm sources of fertility is derivative of cheap oil. But minerals essential to farm fertility can be recirculated within farms or at least within local food systems, and recirculation capacity will become essential to sustainable design.

On-farm recycling. Building high levels of soil organic matter (SOM) will be central to agroecosystem design because SOM is key to achieving not only fertility goals, but also healthy water and mineral cycles, maximal photosynthetic energy capture and use, and optimal biodiversity. Humid, temperate environment soils are exceptional in their ability to store organic matter. French scientist Andre Voisin demonstrated 50 years ago that pulsed grazing (explained below) on permanent pasture is the fastest soil organic matter building tool that farmers have, at least in temperate climates like ours.[7]

The structural element historically proven to work best in these environments is a grass/ruminant complex. This subsystem works on the principle that manure from a portion of the farm devoted to grazing animals will not only sustain the fertility of their forage land, but generate a surplus that will sustain a smaller acreage of annual crops (Figure 4). It can sustain fertility well enough to have generated numerous historical models around the world. The process was used in lowland northern Europe and New England before the industrial age.[8] Cuban research into its potential demonstrated the effective ratio of forage acreage to support cropland fertility to be 3:1 in that environment. In other words, the ruminant stock subsisting on three acres of forage produced enough manure to sustain both the fertility of the forage land and one acre of cropland. This conceptual model, adapted for environmental differences, provides a basis for system design here. Perhaps the most important design question for our purposes is the ratio of forage to cropland that is sustainable in our environment.

The full soil organic matter building process requires a design focus on three crucial areas of the agroecosystem:

6 Peak Phosphorus: The Sequel to Peak Oil. http://phosphorusfutures .net/index.php?option=com_content&task=view&id=16&Itemid=30

7 André Voisin, Grass Productivity (Washington, D.C.: Island Publishers, 1988). Originally published in 1959.

8 Brian Donahue, The Great Meadow: Farmers and the Land in Colonial Concord (New Haven: Yale University Press, 2004).

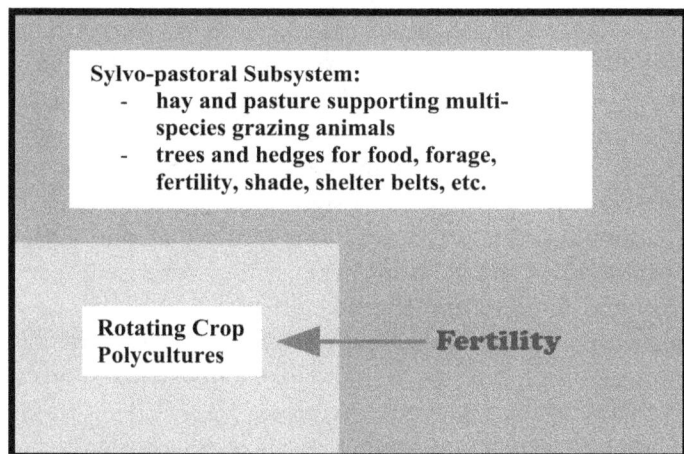

Sylvo-pastoral Subsystem:
- hay and pasture supporting multi-species grazing animals
- trees and hedges for food, forage, fertility, shade, shelter belts, etc.

Rotating Crop Polycultures ← **Fertility**

Figure 4. Fertility subsystem conceptual model

Pasture management for a wide variety of productive, palatable perennial forages, kept in a vegetative state (high growth) by pulsed grazing (see below) throughout the growing season to maximize biomass production;

- Manure storage in a deep litter bedding pack under cover during the cold season to maximize nutrient retention and livestock health;
- Conversion of the bedding pack to compost at a proper C/N ratio during the warm season to maximize organic matter production, nutrient stabilization, and retention;
- Field application of the compost during the warm season as well, to maximize efficient nutrient recycling to the soil.

Pulsed grazing is so important to the success of the soil building subsystem that it warrants an explanation in some detail. Pulsed grazing is a method of repeated grazing of paddocks in a pasture that controls stock density and timing of stock movement in and out of paddocks to maximize forage production over the growing season. This in turn maximizes manure production to build soil organic matter. Forage plants experience repeated pulses of growth and removal of biomass, both above and below ground, over the growing season. Key points :

- Stock enter a paddock before forage leaves its vegetative stage and growth slows.
- Stock leave a paddock while there is still sufficient forage leaf area to jump-start regrowth.
- Grazing causes forage roots to die back, which adds soil organic matter from the dead root mass.
- Stock return to the same paddock when leaf and root regrowth have fully recovered vigor and ability to recover from another grazing.

Recycling from Human Communities. It should be clear from the integrated model (Figure 3) that solving the fertility problem must include repairing the broken nutrient cycle between human excreta and the land. If this seems an insurmountable challenge to modern urbanites, we need only recall from history that whole societies including large cities have managed excellent recycling of "night soil." Among the numerous examples is China, where until the 1950s, 98% of the fertilizer used to grow food came from recycled and organic sources.[9] Relocalization of food production is necessary to reduce the cost of repairing the nutrient cycle. If Tompkins County exports milk products to NYC, what will it cost to return the nutrients in the exported milk to our farmland? In a more county-based food system, methods for recycling humanure and other food garbage that are appropriate to urban, peri-urban, and rural farming sites are more feasible, and will be discussed in the sections devoted to these production systems.

2. Energy Capture

Ancient sunlight in fast-depleting, finite sources (oil, gas, coal) presently supplies over 80% of the energy used in the industrial form of agriculture that produces most of the food consumed in the United States. Natural ecosystems consist of food chains supported entirely by *current* sunlight, so it is easy to design farming systems to work the same way, as was done through most of agricultural history. Solar energy that is accessible directly on farms comes in forms that are far less concentrated than the fossil fuels that we are used to. Therefore we need to design farms that can be productive on far less energy. The challenge is to capture solar energy in as many places as possible as it flows through the agroecosystem.

The carbon cycle is an important way solar energy flows through our world. All metabolic processes in agriculture and other biological systems release carbon to the atmosphere. Tillage that stimulates activity in the soil food web, animal and human digestion and composting are examples. But criticism of these processes as feeding greenhouse gas build-up is mistaken. Biomass conversion to food, fertilizer, or fuel is carbon-neutral over time because its emissions, unlike those of fossil fuels, are part of the biospheric carbon cycle. The important question here is how to manage the carbon cycle to maximize long-term levels of soil carbon sequestered as soil organic matter.

Animal Power. Currently (2009) people tend think of solar capture in terms of relatively high technologies like those that convert wind and sunlight to electricity. Working models exist of homesteads and even farms that are self-sufficient in electricity using small-scale equipment of this sort. However, most analyses of economic viability related to wind/solar electricity production at any scale are based on current costs in the manufacture and maintenance of these systems, all of which still rely on cheap oil. These analyses fail to account for already exponentially rising costs in raw materials and production of the equipment. All production costs of

---

9 http://www.fairviewgardens.org/pub_next_frontier.html

such technologies will rise in parallel with sharply increasing energy costs as the fossil fuel era declines. Like oil, many raw materials used in these technologies are finite resources already on the downside of their historical production curve; they will become unaffordable for many uses in the future. In sum, the window of opportunity that makes these alternative energy technologies approach economic viability now may close in the future as costs begin to rise more sharply. A 10kw wind-electric rig that can power a small farm costs about $70,000, and is usually economically unfeasible even today without subsidies. What will it cost after 15 years of rising manufacturing costs? What will it cost to replace it after its 20-30 year lifetime?

However, there are ways of powering farm production that are more reliably sustainable. Just as the same breeze or brook flowing through a community might be tapped at a number of points for wind or hydropower to run a mill or pump water, solar energy can be captured to produce food or fuel by inserting species appropriately into the farm food chain. Apart from wind and flowing water, solar energy enters the farm ecosystem via photosynthesis in green plants, and flows through the system as one species feeds on another. Large herbivores tap immediately into this chain by feeding on plants that are too fibrous for food use. While they may produce food and fertility as previously described, they will do double duty as work animals in the future, thus replacing no longer affordable fossil-fueled machine labor.

Fields that grow the forages that support work animals and other grazing and foraging species will not compete with cropland. On the contrary, forage fields will provide an essential ecological service as the permanent cover necessary to sustain soil health on all sloping land. Present hillside cropland is always eroding and will be revealed as unsustainable when the crutch of cheap synthetic fertilizer is no longer available. This means that land use plans in hill country like ours will need to include a mosaic of hillside forage land and relatively flat cropland. Unless terraced, the hillsides will be most erosion-free and productive when planned to mimic natural tree-dotted savannas, as hay/pasture that includes fruit and nut orchards, for example. The trees themselves will be multi-functional, producing food or forage, improving the cycling of soil nutrients, providing windbreaks, and shading the grazing animals.[10] Integrated as described here, draft animals like oxen, mules, and horses will optimize the health and productivity of the agroecosystem.

Biofuels. Energy for winter heating and for cooking is almost as important as food production for survival in these latitudes. As much as possible of that energy should come directly from the sun, as in passive solar designs for both heating and cooking. But rural land use will need to reflect increasing local dependence on firewood for the rest. Sustainable forest management and harvest will again become a significant share of rural agricultural production, but serving local urban and village communities not faraway paper mills. Forest conservation and reforestation should start with places that need to be forested for additional reasons, like ridge tops that protect water catchments, and hedgerows that serve as shelterbelts and browse for livestock.

Production of most other biofuels at any significant scale has been criticized as unsustainable on many counts. One that may prove sustainable is small-scale biogas generation on farms, because it extracts methane from some of the farm's normal manure production before it continues in the farm's nutrient cycling loop, as in Figure 3. Most attempts at biogas generation on US farms have been large-scale, high-technology projects aimed at fixing the pollution problem caused by industrial scale dairy farming. So far, farmer adoption of the expensive and complex equipment has been poor, despite subsidies. Meanwhile, small scale biogas generators aimed at producing light and cooking fuel in Third World peasant communities have proliferated, because they cost as little as $30.[11] Biogas production requires no separate biofuel crop that might compete with food production, or inefficient distillation process. For these reasons biogas production at an appropriate scale merits consideration as a way of capturing solar energy as methane fuel for limited use on farms and perhaps even surrounding communities.

3. Water Capture and Use

We live in a climate that is wet yet subject to droughts during the growing season. High productivity food production requires a constant water supply to cover these gaps. Maximizing productivity in the small areas devoted to urban agriculture is especially important, because of their high value in a relocalized food system. Sufficient water falls on urban areas and needs to be conserved there. Barrels can catch only a fraction of roof runoff, and will not be enough for the irrigation needs of a successful urban and peri-urban agriculture. Small water catchment ponds must become a normal part of both the public and residential urban landscape. Pavement runoff will need to be directed to the larger ponds, which might be located in parks and community gardens.

---

10 Karl North, "Optimizing Nutrient Cycles with Trees in Pasture Fields," *LEISA Magazine*, 24 (2), March 2008. http://www.ileia.org /index.php?url=magazine-list.tpl&p[source]=ILEIA

11 T. R. Preston, "Biodigesters in Ecological Farming Systems," *LEISA Magazine*, 21 (1), March 2005. Also http://www.ruralcostarica .com/biodigester.html

Rural agriculture will need more extensive water capture plans to hold and use water for farms and whole watersheds. Such a system should be gravity feed system, in order to avoid the increasingly high cost of pumping. An example is the keyline plan that traps some surface water in upper fields and directs the excess into strategically located irrigation ponds.[12]

Our irrigation needs in New York may be intermittent but still will require a lot of pipe and other delivery hardware when scaled up to cover all food production land. Rising costs of current irrigation delivery systems may become a limiting factor, forcing the invention of ones that use cheaper materials. This has been the experience in Cuba, whose year-round agriculture is heavily dependent on irrigation. Cuba's artificially triggered "peak oil" experience has been a bellwether and a source of lessons for the rest of the world.

Ponds will be needed to serve numerous purposes, as in Figure 3. Basins to process biodigester outflow and other organic liquid waste can grow algae and duckweed for animal feed, and then feed the cleansed water into ponds for fish and other aquaculture, as in Figure 5. They will attract aquatic life including species useful for garden pest control, and enhance human quality of life as they beautify places and improve microclimates.

Wetlands abound in New York and are among the most productive natural ecosystems. Because of their natural potential, they can be harnessed for highly productive agricultural use yet be managed to retain much of their natural function. Historical and contemporary models include wetland systems that fed older civilizations from the Aztecs to the Incas in Latin America, as well as many parts of Southeast Asia today. Typically, as

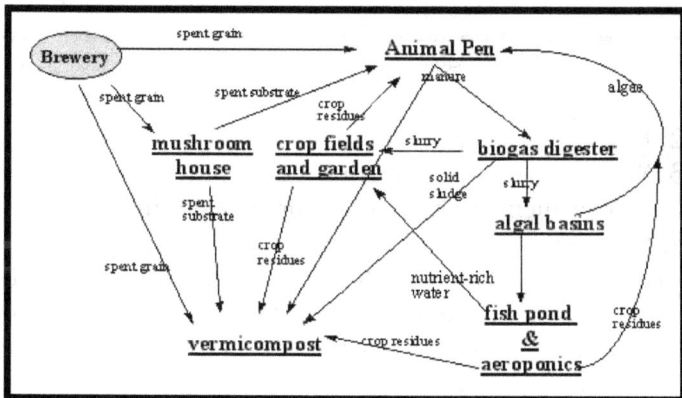

Figure 5. Facilities for bioconversion using the UNU/IAS integrated biosystem at Montfort Boys Town, Suva, Fiji

in the Aztecan systems known as chinampas, farmers cut canals through the wetland and use the soil to create beds raised above the water level for agricultural use. The canal system is designed to allow the water control that keeps the raised beds well watered without being subject to undesirable flooding. Because of the ubiquitous water, these wetlands are highly productive as both agricultural and aquacultural systems. They produce so much biomass that they tend to maintain their own fertility, dredged from the decomposing detritus in canal bottoms.

One such wetland, adapted from lowland English agriculture, became the core of a highly sustainable agricultural system that supported the population of colonial Concord, Massachusetts for many generations.[13] The Great Meadow that traversed the village and all other nearby riverine flood plains was a swamp commons that was first flooded to deposit silt, then partly drained and reserved for pasture and hay as it dried out during the growing season. As in parts of Europe, these well-watered riverine meadows produced enough livestock feed, livestock, and manure to sustain the fertility of the adjacent dry lands devoted to tillage agriculture. Figure 6 shows that already by 1650 careful allocation of land use had taken place on a functional level to sustain the whole system. Historical models like these suggest that we will want to regard modified wetlands as an important agricultural asset in the energy descent era.

4. Pest Control

From a systems perspective, pest problems are "structural," hence best addressed by system design rather than treatment with pesticides. In this section I will summarize two main strategies addressed in order of importance: a focus on the food species themselves, and then the layout of the physical and biological environment as it affects these food species.

Much as health care in humans requires preventive medicine, we must grow healthy plant and animal species as a first step in pest control. A primary structural problem is the genetic industrialization of most agricultural plant and animal species, which was gradually achieved in modern times by breeding processes that prioritized productivity and short-term profit over other genetic traits, like hardiness. Moreover, relying on pesticides, even "natural" ones, to protect these weakened subspecies inevitably fails over time because pests gradually adapt to conditions and treatments that become heavy-handed and routine. An example is parasite resistance in sheep, which has been neglected and lost. The resulting industrial breeds must be medicated so often that the parasites are gradually becoming immune to most medications. To be sustainable, food production systems will need to return to varieties and breeds that, while sometimes less productive, have more genetic defenses. By genetic selection farmers can rebuild hardiness in industrial breeds as well.

---

12 http://www.keyline.com.au/ad1ans.htm

13 Donahue, *op. cit.*

Recourse to medicinals and other treatments is a strategy of last resort, indicating a design failure in the production system, which must be addressed.

Agriculture and land use

From the foregoing it seems clear that life after fossil fuels will demand much reorganization of food production. To create a local agriculture that feeds the county, the map of rural and urban land use will change dramatically. In the countryside, wetlands and floodplains, hillsides, flatlands, and woodlands will have specific uses designed to maximize while sustaining the productivity of whole agroecosystems. Essential rural land use components might be:

1. Hillsides in forage land sufficient to support cropland fertility.
2. Flatlands in crop rotations.
3. Wetlands and floodplains development and water management for high forage or crop production.
4. Sufficient forest for county firewood and basic construction needs, managed for maximum regenerative capacity, which requires fencing out livestock. Woodland regenerative capacity equaling 1 cord/acre/year is a common rule of thumb.

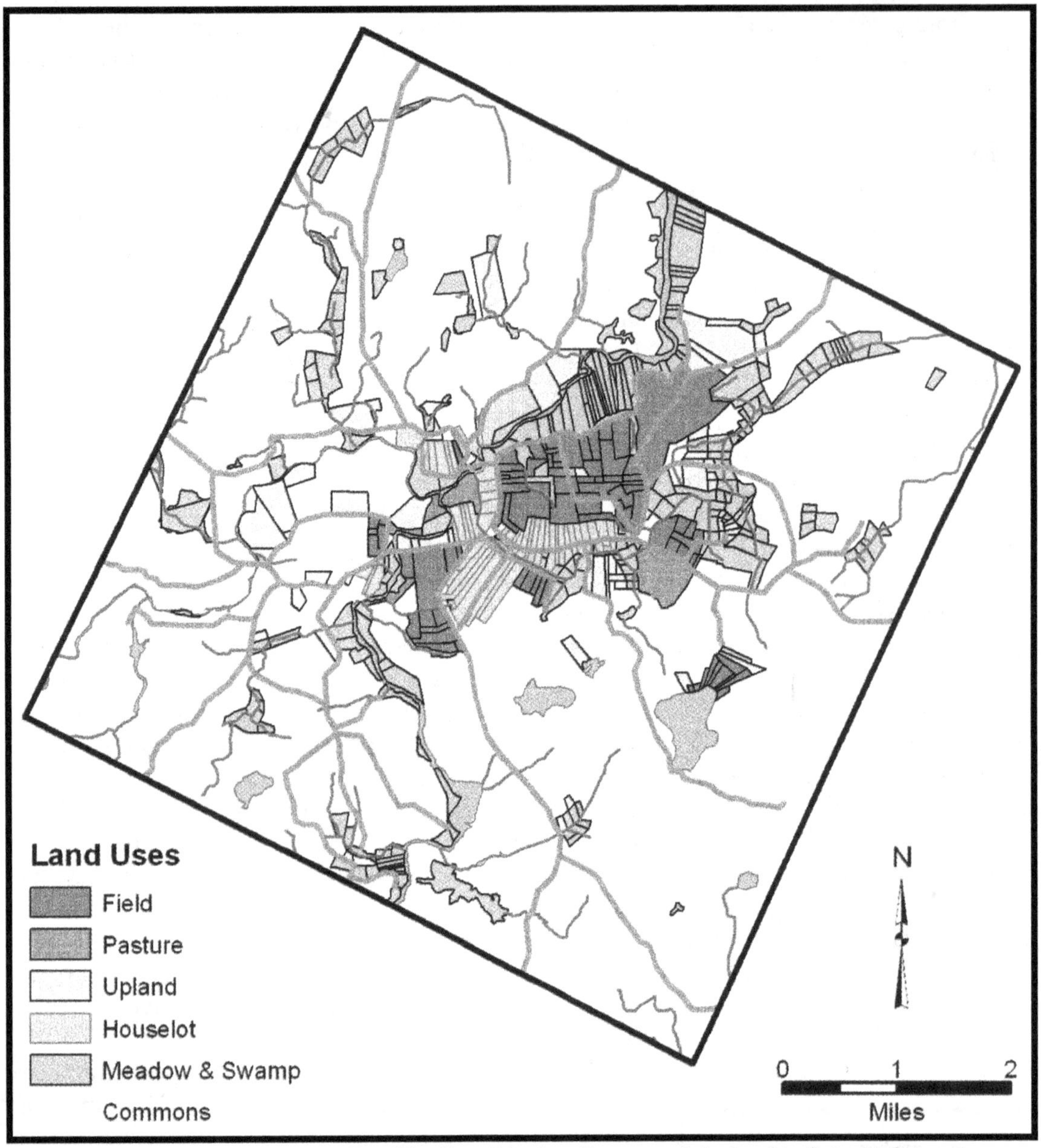
Land Uses
Field
Pasture
Upland
Houselot
Meadow & Swamp
Commons
N
0
1
2
Miles

Figure 6. Concord, Massachusetts, 1652. From The Great Meadow: Farmers and the Land in Colonial Concord

The design of alternative environments uses three general strategies of pest control: luring or driving them away with trap or repellent species or physical barriers; creating species and habitats that attract "beneficials," species that prey on pests; and continually altering the environment with crop and animal rotations that shift them away from pests.

This last strategy points up a characteristic of the natural world that needs to be taken into account: it is always evolving. In the long run this means that pest control strategies can never be permanent, but must always be evolving to stay a step ahead of pests as the latter adapt to current controls. The downfall of industrial pest controls is their heavy- handed strategy of total pest elimination and routine medication. Ironically this creates the environments most conducive to genetic evolution in pest organisms toward immunity from controls.

Many uses of city land will no longer be economical in the coming years. Land will need to be converted to food production and its supporting functions, like composting and water conservation. Prime candidates for conversion are the commercial strips now inhabited by national corporate chain stores. Private and public parking lots, which energy descent writer William Kunstler sees as soon-to-be-dysfunctional "missing teeth in the urban fabric," are another example. During Cuba's artificially triggered encounter with "peak oil," public interest dictated that a better use of resources was to raze ageing buildings to create urban garden space, rather than to restore them.

In the integrated system approach described here, the functions of plants and animals will undergo marked changes. The functions of many species to facilitate tight nutrient cycling, labor, and other services that underpin the health of the whole agroecosystem, will become more important. In the case of some animals, these functions will become primary, and food production will become a secondary function, with numbers of animals on farms directed to their primary functions. The result will be a general production system model that aims for maximum sustainability, remains within the carrying capacity of the natural resource base, and within that framework, feeds the maximum number of people per acre of land used.

SEEING COUNTY FOOD PRODUCTION AS AN INTEGRATED WHOLE

I said earlier that providing for the local food needs of urban populations requires a design that integrates three overlapping categories of production systems: *urban agriculture systems* (many small islands of gardening in the city center), *peri-urban agriculture* (larger production areas on the immediate periphery), and *rural agriculture* (feeder farms associated with village-size population clusters in the hinterland of the city but close enough to be satellite hamlets). We'll shortly be looking at each of those systems in detail and considering some possible scenarios using locations in Tompkins County as cases in point, but first I'll try to picture the future county food system as a whole: its historical context and implications, and interdependencies among the parts that will make them most effective as an integrated system.

Learning from history: pre-fossil fuel food miles

How relocalized does a food economy need to be in the energy descent era? Throughout history, food security everywhere has been heavily dependent on a reliable supply of staple foods, especially starch staples like root crops, pulses (beans, peas, etc.) and grains. Our region once was self-sufficient in staples but gradually imported most of them. To regain food security, we must establish a measure of food sovereignty as local policy, especially in staple foods.

A look at NYS history is a reminder that easily conserved and transportable food commodities traveled far before the railroads existed, and to a degree even before the canal system was built.

Pre-canal overland commerce in high-value imports and industrial goods, paid for in farm products, was common across New York State. The account in Figure 7 shows the sorts of goods that flowed in both directions.[14]

The general store found in every hamlet undertook the task of supplying to its people food, drink, raiment, tools, and all the miscellany of living that could not be produced on a farm.

1st January, 1822.

More New Goods.

THE subscriber is now receiving a large addition to his Stock of MERCHANDISE, which makes his assortment very complete, which will be sold

Cheap for Cash,

OR EXCHANGED FOR

Pot and Pearl Ashes, Whiskey, Wheat, Pork, Lard, Butter, Rye, Corn, Oats, Flax, Timothy Seed, Clover Seed, Bees Wax, Tallow, &c.

CASH PAID

For most of the above named Articles.

The storekeeper furnished to the housewife tea, coffee, sugar, spices, salt, and such luxuries as the times afforded. The store carried ginghams, muslins, silks, calicoes, combs, beads of women's wearing apparel, and for the men, broadcloth, denims, homespuns, vest patterns, and coat patterns, for rich and poor alike. The carpenter purchased nails and edged tools; the tailor his thread, buttons, and shears; and the shoemaker his thread, wax, awls, and sometimes leather. In the store the sale of spirituous liquors was conducted by the measure in amazing quantities. Of cash to pay the storekeeper there was precious little, and most of the business was done by way of trade. The farmer exchanged for the storekeeper's wares grains and seeds in assortment; butter and eggs; meats, pelts, hides, tallow, and lard; often the housewife turned in the product of the spindle and loom for the goods she needed.

Figure 7. Goods that historically made up the bulk of commercial trade in 19th century rural New York

By 1830 the New York canal system linked most agricultural depots of the state to waterways--the Great Lakes and lesser lakes like Lake Champlain and the Finger Lakes to the main state rivers--and thence to the population centers and to foreign trade. Figure 8 is an account of the primary commodities in the lake traffic through Buffalo in 1847 and provides a rough measure of the tonnage and kinds of foods that moved long distances in that era.[15]

Flour, bbls........	1,857,000	Oats, bu.........	446,000
Pork, bbls........	42,000	Butter, kgs.......	101,584
Beef, bbls........	38,900	Lard, lbs.........	3,436,000
Staves, ps........	8,800,000	Cheese, bxs.....	30,840
Wheat, bu.......	6,489,100	Cheese, casks.....	6,450
Corn, bu.........	2,862,000	Lumber, M.Ft.....	17,313

Figure 8. Great Lakes traffic arriving at Buffalo, 1847

In the late 19th century the railroads took over most transport of farm products out of rural areas; even certain bulkier items that travel well like potatoes, onions, cabbage, and livestock were included in state-wide commerce and beyond.

---

14 Ulysses Prentis Hedrick, *A History of Agriculture in the State of New York* (Albany: New York State Agricultural Society, 1933).
15 Ibid.

Apart from food security, the stimulus to the local economy and the provision of fresh, superior quality food are good reasons to produce as much food locally as possible. But consideration of the above historical perspective suggests that the question of *how much we need to depend on locally produced food* turns on the ability of the state to promote the revival of the railroads or, failing that, at least the canal system. The existence of long-distance trade before the era of *energy ascent* in products like grain that travel well suggests that during *energy descent* widespread trade in some agricultural products will persist despite rising transport costs.

However, many energy descent analysts[16,17,18] believe that the US economy has been so undermined by internal and external debt and dependence on fossil fuels that state and federal institutions will eventually be unable to maintain the present social order, much less take on the reconstruction of pre-oil transportation networks. This scenario suggests the need for a high level of local food production. Analysis of probable futures at this macro-level clearly suffers from the uncertainty surrounding so many of the key variables. Perhaps the best insight one can draw from the records of earlier food systems is a ranking of agricultural products for localization, according to their sensitivity to a shrunken distance economy.

Even assuming the construction or restoration of energy-efficient transport networks, other concerns ultimately will force increasing dependence on locally grown food. A sustainable food system must recycle nutrients. The historical expansion of US food miles relied first on the depletion of fertile virgin soils, then on cheap fertilizer and other manufactured inputs. Without the crutch of increasingly expensive inputs, declining agricultural yields in farms distant from consumers will force large foodsheds to shrink over time. Even proposals for the reorganization of the national and global food system into bioregional systems or foodsheds larger than counties have ignored the nutrient cycling imperative, which becomes increasingly difficult as food is grown farther from where it is eaten. This raises the question of how to feed large cities in a purported Northeast foodshed and still sustain the health of the soil that grows the food.

As early as 1862, scientists were writing of a *metabolic rift* that had developed between city and countryside.[19]

The rift was both biological and social; the nutrient cycle had broken as the nutrients that fed urbanites no longer returned to the rural lands where the food was grown, and urbanites had lost appreciation of the fact that urban prosperity ultimately depends on the health of the land and its natural systems.

The social/cultural rift may be the biggest obstacle to change. The very existence of cities depends on the accumulation of a surplus of wealth from agriculture and other raw material extraction from the land. The temporary ability of humanity to substitute fossil fuel dependent technologies for human labor and the soil fertility and other services originally supplied by natural systems created the illusion that human labor and ecological services are of little importance in agriculture, and therefore have little bearing on the question of the survival of cities. Technology, apparently an urban product, became paramount in the hierarchy of urban cultural values. In that hierarchy, technology could even replace the social capital of healthy families and communities that traditionally gave agrarian society much of its strength and resilience.

The county needs to be ready for these challenges. The limiting factor that inhibits food system change is not biophysical knowledge of how to do it, but social knowledge of the power structures that have closed down local food economies and prevented their revival. Successful strategies for change can emerge only from a deeper understanding of how things work in the system of power relations, both in the county and beyond.

A county policy framework that effectively favors local production and reverses the power shift in modern society toward centers that today exploit peripheries will ultimately improve local quality of life. In the early 19th century, before the rise of competition from the Midwest, agrarian NY communities sold to nearby cities and enjoyed a relative prosperity that reflects the true dependence of urban affluence on the wealth of the land. Recently it was estimated that in Maine, $10 a week spent on locally produced food would put $104 million into the state's economy.[20] This suggests that a public program to relocalize the county food economy eventually could sell politically as a core element in regenerating the local economy overall.

16 Chris Martenson, http://www.chrismartenson.com/crashcourse

17 Richard Heinberg, *Peak Everything: Waking Up to a Century of Declines* (Gabriola, BC : New Society Publishers, 2007).

18 James Howard Kunstler, *The Long Emergency* (New York : Atlantic Monthly Press, 2005).

19 The earliest author to apply the term *metabolic rift* to the "robbery" of country soils through the exportation of food to cities appears to have been the German chemist Justus von Liebig in the introduction to the seventh edition of his *Organic Chemistry in its Application to*

Agriculture and Physiology. The term was later used by Karl Marx and others. See J. B. Foster, "Marx's ecology in historical perspective," http://pubs.socialistreviewindex.org.uk/isj96/foster.htm and Rebecca Clausen, "Healing the Rift: Metabolic Restoration in Cuban Agriculture," *Monthly Review,* May 2007.

20 Community Food Security Coalition, "Urban Agriculture and Community Food Service in the United States: Farming from the City Center to the Urban Fringe," FoodSecurity.org, October 2003. http://www.foodsecurity.org/PrimerCFSCUAC.pdf

Interdependencies in the county food system

The three types of county agriculture to be explored in this series are best suited to different, complementary roles in county food production. Taking its cue from the pattern in earlier times, urban agriculture will give priority to production of vegetables and fruits for fresh consumption that can be grown intensively, in raised beds for example. Peri-urban agriculture will supplement urban gardens with produce that requires more space, and will support some livestock. Rural agriculture will be responsible for most of the large animal production and large-scale field cropping. A high priority of farming in satellite villages will be to grow the bulk of the staples, like potatoes, oats, roots, brassicas, legumes, squash, alliums, and apples, which have proven to be dependable in cool, temperate climates. The county will need to rely mainly on outlying farms for nonfood essentials as well, such as oilseeds, flax, hemp, wool, leather, and wood.

Because the agriculture of the future will need closed nutrient cycles, fertility for all county food production cannot be considered apart from county organic waste streams.[21] To maintain fertility, organic waste must return in some form to food production sites. As the dense urban population produces the bulk of the waste, public institutions will need to take responsibility for separate collection of the purely organic component of the urban garbage and sewage waste streams, recycling part of it back to rural farms.[22]

Fertility in urban and peripheral agricultural soils can be sustained with compost from the city organic garbage stream alone. A study of one urban community revealed that urban agriculture alone could absorb 20% of the

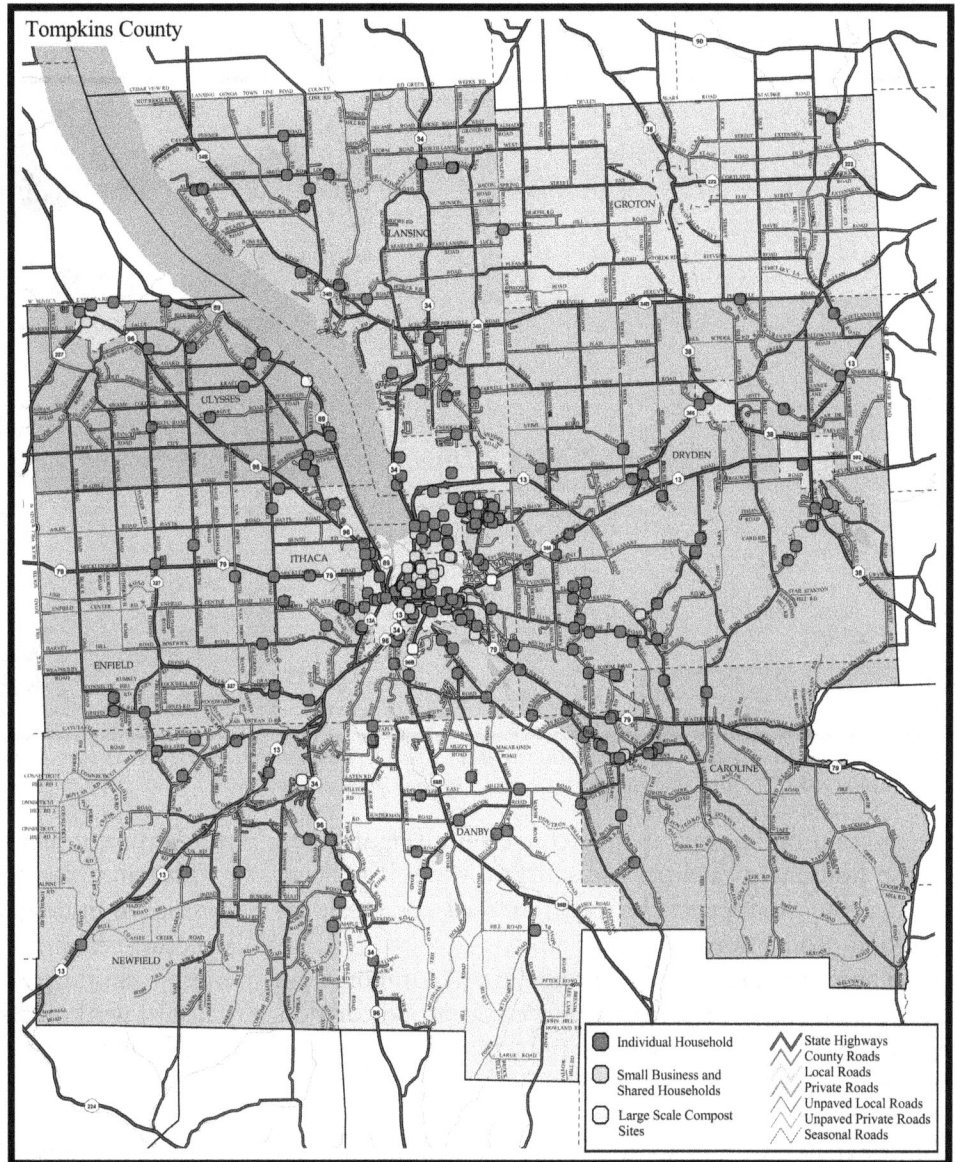

Figure 9. Composting sites in Tompkins County

organic waste production of the city.[23] This will require a municipal policy and program of careful triage, collection, and composting at optimum C/N ratio by mixing high-nitrogen food garbage with high-carbon sources like leaves and shredded paper trash. The city could assign responsibility to urban institutional sources, such as schools and restaurants for moving their large organic waste streams to composting facilities at specific peri-urban food production sites. A map of existing Tompkins County composting sites demonstrates the composting potential (Figure 9).[24]

---

21 For information about local waste processing facilities, see the TCLocal article "Wasting in the Energy Descent: An Outline for the Future" by Tom Shelley, online at http://tclocal.org/2009/01/wasting_in_the_energy_descent.html and in print in *Thinking Local in Tompkins County Vol. 1*.

22 Tom Shelley has recently begun to prototype this process with "The Sustainable Chicken Project," which returns nutrients to the land by collecting kitchen scraps in the City of Ithaca on a subscription basis and feeding them to chickens at Steep Hollow Farm three miles outside the city in the Town of Ithaca. See http://www.sundancechannel.com/sunfiltered/2010/01/sustainable-chicken-project/ and the farm's blog at http://steephollowfarm.wordpress.com/

23 Luc J. A. Mougeot, *Growing Better Cities: Urban Agriculture for Sustainable Development* (Ottawa: International Development Centre, 2006). http://www.idrc.ca/openebooks/226-0/

24 http://www.co.tompkins.ny.us/gis/maps/pdfs/CompostMap2000-E.pdf

As for sewage, eventually Ithaca will have to desewer, converting to urban night soil collection, biogas extraction, and the recycling of residual organic matter to county farms that will be necessary to maintain the mineral content of rural agricultural soils. In the short run, guerrilla humanure composting from backyard compost

sources that will shape the design of urban agriculture systems in the city of Ithaca, and then offer a case study as a design example.

The high institutional and population density of urban areas promotes labor-intensive production methods, community regeneration through cooperative management, and transport efficiency for agricultural inputs and products. The ability to have more farmers per acre permits the kind of management-intensive system that maximizes productivity through close monitoring and good timing throughout the growing season. Increased headcount allows a division of labor to manage diversified production integrated into one system. One neighbor could grow rabbits (Figure 11) and provide manure and meat while another grows vegetables and a

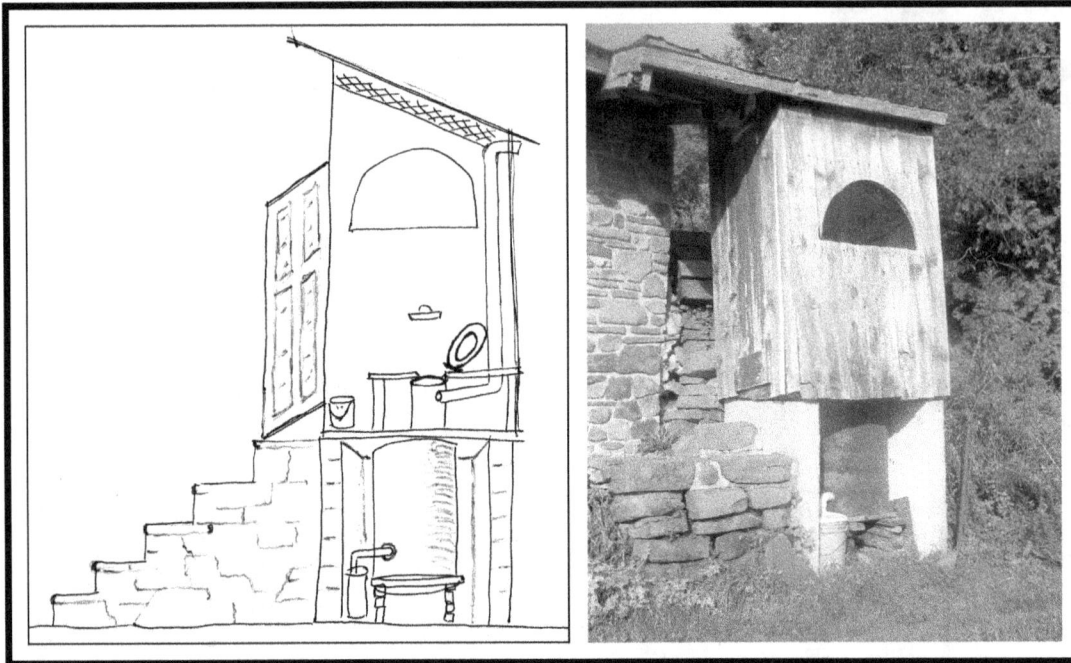

Figure 10. A functioning home-built composting toilet based on a 55 gallon drum that has been in operation in Cortland County since 1983. The drum is periodically rotated out through a composting cycle

toilets can build toward full conversion (Figure 10). These household facilities are satisfactorily self-policed, because the product will be used in closed-cycle residential food production.

To sum up: some of the current massive importation of the county's food consumption could go on for decades. I've pointed out serious risks to food security if this were allowed to continue and argued that the distance economy in food causes metabolic rifts that make it ultimately unsustainable. A local food production system can mend the biological rift, and the reorganization of county agriculture itself will begin to address the most challenging rift, the social and cultural rift between urban and rural life.

URBAN AGRICULTURE
Providing for the local food needs of urban populations requires a design that integrates three overlapping categories of production systems: *urban agriculture systems* (many small islands of gardening in the city center), *peri-urban agriculture* (larger production areas on the immediate periphery), and *rural agriculture* (feeder farms associated with village-size population clusters in the hinterland of the city but close enough to be satellite hamlets). I'll now turn my attention to each of these systems in turn, beginning with a look at the needs and re-

third concentrates on fruits.

The abundance of city institutions presents opportunities to build gardening appendages on existing social structures organized for other purposes. In the sudden energy shortage that transformed Cuba's agriculture, schools, workplaces, and even governmental institutions were quick to become partly self-sufficient in food production. As awareness builds that gardening is a form of physical education whose value increases relative to, say, football, schools will see the need to devote more playground space to school gardens.

Figure 11. Urban rabbit hutches in Cuba

Intensive Design

The high productivity of urban agriculture has proven itself in many cities, notably in the severe food crisis that Cuban cities experienced in the 1990s.[25] Productivity in urban agriculture comes in great part from intensive design and management. The greater labor required for intensive production is potentially available in urban agriculture and can make it highly productive in several ways. Space can be used more efficiently than in extensive row cropping. Intensive growers can plant many vegetables in permanent beds instead of rows, minimizing walk or machine alleys between rows and concentrating soil building in the beds rather than the whole field. Also, farmers can plant crops of fast maturing foods, like salad or cooking greens, in spaces between large, slower maturing ones like broccoli. This practice of planting so-called catch crops makes more intensive use of limited space during the growing season. Tiered design that uses light efficiently is possible. Crops can be grown in companion polycultures to trade ecological services; legumes like pole beans fixing nitrogen for the corn that provides the pole, or a row of peas climbing a wall while fertilizing a row of carrots. Maximum use and close management of protective devices like frames and cloches permit not only season extension but also more effective temperature and moisture control of plant growth during the regular season. Finally, the consumers of urban-grown food are close enough to permit effective recycling of nutrients into the garden soil via backyard compost piles and composting toilets, partially or totally eliminating the need for space for compost crops.

For these reasons, urban spaces can be nearly 15 times more productive than rural farms.[26] In World War II, residential "Victory gardens" in the US produced a quantity of fresh vegetables equal to the total commercial output of these foods.

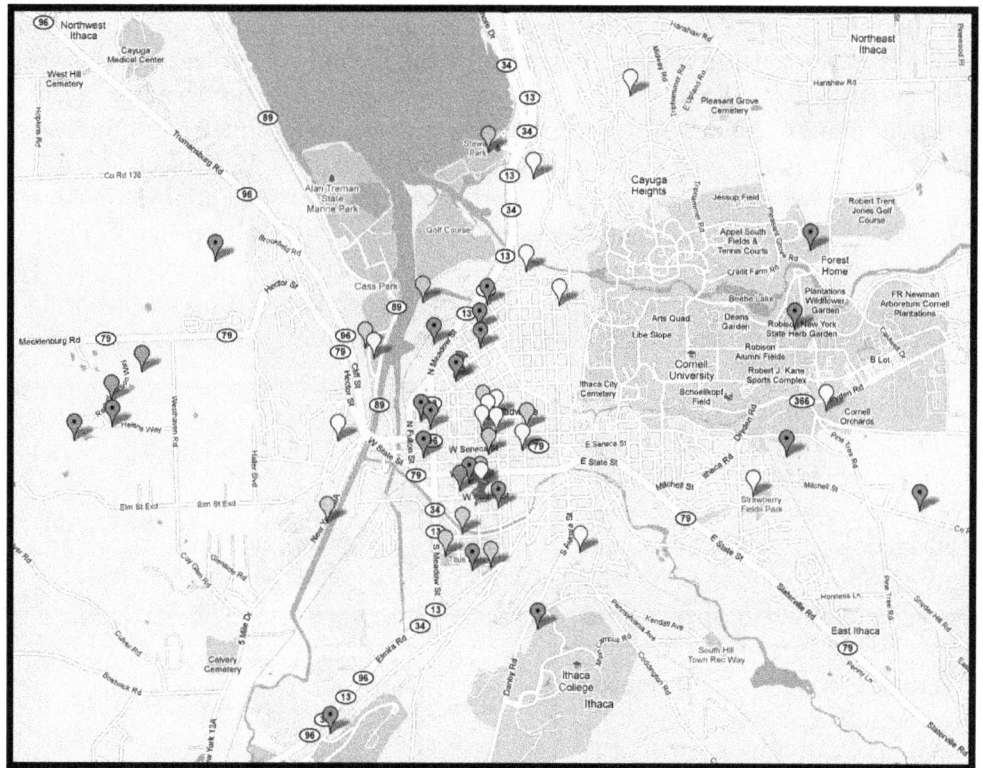

Figure 12. Community and School Gardens of Tompkins County. Color version (see link in footnote) distinguishes community gardens, school and educational gardens, farmers' markets, and sites that have expressed an interest in gardens or have been identified as potential sites for new community gardens

The Ithaca Urban Environment

Ithaca's topography of central flatlands surrounded by steep hills presents distinctive opportunities and constraints for urban garden design in each area. Josh Dolan's map of current and potential community and school garden sites in Tompkins County illustrates some of the possibilities.[27]

On the hillsides, some food production will require terracing, but the many south and west facing retaining walls and house walls in residential neighborhoods on Ithaca's steep hills provide opportunities for vertical growing. This will maximize use of space, which is important in urban gardens. Vine plants can sometimes grow either from the top of the wall down or from the bottom up. Twine or poles laid against the walls help plants like tomatoes and beans get a grip going up, and planks or slates shoved between wall stones support heavy fruits like melons or squash as they grow bigger. Projections of climate change for the Northeast include a 20 to 30 percent increase in winter precipitation over this century, but hotter summers when water is needed for growing, suggesting a greater need for seasonal water capture.[28]

25 Catherine Murphy, *Cultivating Havana: Urban Agriculture and Food Security in the Years of Crisis,* Development Report Number 12 (Food First: Institute for Food and Development Policy). http://www.foodfirst.org/pubs/devreps/dr12.pdf

26 Michael Ableman, "Agriculture's Next Frontier: How Urban Farms Could Feed the World" (Center for Urban Agriculture at Fairview Gardens, 2007). http://www.fairviewgardens.org/pub_next_frontier.html

27 http://maps.google.com/maps/ms?hl=en&ie=UTF8&msa=0&msid=11296740563107444443966.00046b4b4eb5e29a3ab69&t=h&ll=42.435707,-76.459758&spn=0.014475,0.026994&z=15

28 *Confronting Climate Change in the Northeast* (summary of a 2007 study conducted in part by the Union of Concerned Scientists). http://www.climatechoices.org/assets/documents/climatechoices/new-york_necia.pdf

The hills of Ithaca have great potential for gravity irrigation if water is distributed downhill through many residential gardens. Pools at each site can store water to provide gravity irrigation to terraces via berms and swales. Institutional sites might justify tapping this gravity flow to power small grain mills or electric generators. On the city's flatlands, current uses of many commercial sites will become obsolete in the energy descent. Energy inefficient businesses and parking lots will become prime sites for takeover by guerrilla gardeners, building pressure for legalization. Water is relatively abundant in our environment, but because of its importance for highly productive food growing, water reserves collected from roof drains into garden-side irrigation pools will be vital to build resilience into urban production systems.[29] More resilience can be achieved by routing roof water into attic or upper story tanks for household use and then channeling the overflow into irrigation pools.

Visioning an urban agriculture case

A group of neighbors has decided to form a loose gardening cooperative, because a pooled effort will solve the core production problems of fertility, water, pest control, and energy more efficiently than would completely individual projects as well as promoting the sharing of equipment and pooling of knowledge. In individual backyards they have been growing a few vegetables and fruits, often in containers they can bring inside for extended season growing.[30] Many neighbors have enough small stock such as rabbits, chickens, and pigeons to process organic kitchen garbage; however, their yards are mostly too small for the amount of food they want to produce as a co-op.

The neighborhood group has agreed to devote most backyard space to compost production and the collection of irrigation water for the co-op. They have quietly attached composting toilets to their houses and built filter/digesters for household greywater and little ponds to store greywater and roof water, while currying support for legalization when the time is politically ripe. Eventually the city created property ownership and lease contracts with management agreements that provide incentives for ecological management, like composting of residential waste streams and maintenance of food perennials on the property.

To make space for the main garden the neighborhood co-op razed a building abandoned as too costly to renovate for energy efficiency, and depaved an adjacent parking lot that became obsolete when the city got serious about public transportation. The land owners were happy to lend the properties in long-term agreements because the city had created land tax credits for land lent for urban agriculture. As in the urbanization of agriculture in Cuba (Figure 13), our neighborhood co-op often left rubble in place and created raised beds over it with soil imported from nearby rural farms and compost from backyard and municipal production sites. This photo also illustrates the use of a pest insect trap crop of corn planted at the end of the raised beds containing other crops.

Figure 13. Urban coop garden, Pinar del Rio, Cuba

The co-op employs a master gardener to design and manage the garden to include the polycultures, rotations of crops among beds, water, compost, and mulch acquisition and application that will maximize the health of the system. Because it integrates a greater diversity of crops and habitats, this system achieves a higher level of sustainability than community gardening by individual allotment. Each household is assigned responsibility for working a section of the garden under the direction of the manager. As different crops or polyculture combinations rotate through each section, all neighbors gradually have become skilled at growing all the foods that the co-op produces. The manager arranges for extra labor when necessary, as in planting and harvesting, for compost and water from backyard ponds, and for supplemental compost from the city's public composting enterprise.

The project design includes a number of elements not yet found in many urban gardens: hot and cold frames and nursery beds to feed transplants into the garden; glass bed covers to provide season extension; habitats for

29 Two resources on water management for urban agricultural use: ftp://ftp.fao.org/docrep/FAO/011/ak003e/ak003e05.pdf; http://www.ruaf.org/sites/default/files/Chapter%209.pdf

30 http://www.gardeningknowhow.com/urban/designing-your -container-vegetable-garden.htm

beneficials and other native species; insectaries, bird houses and trap and repellent crops for pest control; border hedges of nut and fruit bushes and trees and other perennial crops; and artistic corners in which to rest and enjoy the garden.

The neighborhood co-op provides regular shares of harvests to its members, and sells surplus produce in a market stand on site using the local county currency. Some members operate small processing enterprises to preserve co-op output for the neighborhood.

This model of urban agriculture may work in a number of locations, but many other models will be needed that are adapted to conditions of specific sites or parts of the city.

Figure 14. Cooperative farms on the edge of Havana, Cuba

PERI-URBAN AGRICULTURE

Moving out from the urban center with its many small gardens, the second of the county food systems uses larger production areas on the immediate periphery. As in the case of the urban system, I will offer a case study of peri-urban agriculture as a design example.

Cities are often ringed with suburbs, parks, and industrial and commercial zones that can be converted to larger, more integrated agricultural systems than densely populated urban neighborhoods (Figure 14). Deer and rodents have proliferated in the urban-suburban boundaries that are excellent edge habitats for these species. Agriculture in these areas will need to achieve deer and rodent control by fencing that is effective against jumping and burrowing and by regulated trapping for meat and hides to eventually reduce populations.

The best candidates for conversion to farming are sites that have good soil and water resources yet are close enough for easy access by urban consumers and potential farm labor. Two such areas on the periphery of Ithaca are the flood plain beside the lake and inlet and the nearest locations on the main existing transport routes, particularly those with existing rail lines, north up the east edge of the lake and south along route 13.

The flood plain

One-sixth of 19th-century Paris was devoted to intensive urban gardens, prominently in the Marais (wetland) on the right bank of the Seine River. Fueled by manure from the city's thousands of working horses, peri-urban gardens fed Parisians with greens, vegetables, and fruits the year around. The history of a similar district on the edge of climatically similar Ithaca indicates its food production potential. This neighborhood was once home to a distinctive waterside community of fisher-farmers who, despite their lower socio-economic status compared to some Ithacans, were able to achieve relative self-sufficiency on the rich alluvial soils and aquatic resources of their neighborhood.

Ithaca has a unique resource in these lakeside and inlet soils. They are potentially the most productive agricultural land in the county when converted to the chinampa-style systems described earlier (see Figure 15).

Some of this land may now be "brown fields" of soils that are polluted from years of commercial and industrial use but can be reclaimed biologically. Bioremediation can take various forms. Several years of intensive grazing and repeated trash plowing and replanting of grass cover not only builds soil organic matter rapidly but cleanses it as well by bacterial action as the soils become more biologically active. Instead of normal plowing that buries sod, trash plowing upends it for fast aerobic decomposition. If this is insufficient, raised beds with imported soil are a solution that has worked in many urban locations.

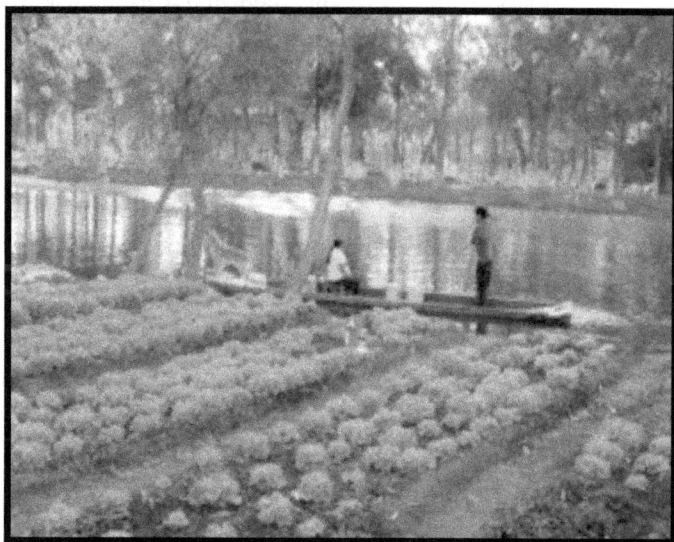

Figure 15. Mexico City chinampas

Land use policy for the district would have to change to reflect the food production priorities of the energy descent. Some lands now dedicated to industry, the commercial strip of big box stores, and parts of parks and the golf course will be acknowledged as prime farm-

Figure 16. Examples of potential waterside farm sites on the edge of Ithaca

million today.[31] In a similar assessment, Swedish systems ecologist Folke Günther estimates that the rural farming population needed to support an urban community should be 12 times the urban population. The starting point in our case is a county population of 100,000, of which 30,000 is urban. To achieve the necessary balance, Günther suggests relocation of some urban and close suburban populations to clustered housing in satellite farming villages[32] as older urban housing is replaced by urban gardens. The most economical location for some of these peripheral ecovillages might be in the peri-urban agricultural district along the main transport routes near the city.

Ideally this process would be part of a general physical redesign of both the urban and hinterland communities according to the model that emerged in Europe, where centuries of higher population densities have dictated more careful land use planning. Even today, European towns large and small are characteristically dense clusters of buildings that end abruptly in agrarian vistas.

Visioning a peri-urban case: Waterside Gardens
Commercial strips and malls that typify the urban edge, vacated in the shrinking national economy, are prime candidates for a public takeover that would convert their parking lots to agriculture and the empty buildings to farming and related community uses. To exemplify this conversion, we will envision a farm operated as a commercial cooperative, using a future abandoned Wegmans waterside parking lot and supermarket building (one of the locations outlined in Figure 16). Let's call our imaginary cooperative "Waterside Gardens" (Figure 17).

land. Figure 16 illustrates examples of potential waterside farm sites.

The politics of conversion of water-side lands to prime food production sites will require a new mindset. Agriculture may be the best use of some of the land now devoted to recreational activities like sailing, picnicking, and golf. Consumers accustomed to shopping in national chain stores will need to learn that they represent what Wendell Berry in The *Unsettling of America* called an extractive, colonial economy. This economy transfers wealth to metropolitan centers of power from rural peripheries and operates at many scales, from impoverished banana republics like Nicaragua, to shrunken agricultural towns in Nebraska, to the depressed areas of upstate New York. Thus the national chain stores that ring the Ithaca periphery are economic "monocultures" that strip economic wealth from the county just as agricultural monocultures drain fertility from the soil.

Transport route locations
Conversion to more sustainable food production requires more people living closer to food production in order to provide labor and to facilitate nutrient recycling. Energy descent writer Richard Heinberg estimates the need for 50 million farmers in the U.S., up from 2

31 http://www.energybulletin.net/node/22584
32 http://www.holon.se/folke/lectures/Ruralisation-filer
/v3_document.htm

A policy framework. The dirty little secret of small farms is that they don't make much of a profit in competition with industrialized agriculture. A food policy framework guarantees the economic viability of Waterside Gardens:

- As part of a county-wide green belt policy to stop and convert urban sprawl, the city has re-municipalized most of the inlet area from the lake front to Buttermilk Falls, providing a free lease to co-ops like this one as long as they continue to build food security and food sovereignty in the county.
- In the wake of widespread demand for local food sovereignty, the country has revised the Constitution. As part of a growing reliance on local, county-wide economic policy making, a tariff is now levied on food coming into the county based on food miles and the ability of local agriculture to provide the product.
- A trolley stop on the public light rail line serves the site to bring agricultural inputs to the co-op and consumers to its retail food market.

Models of ecological health and productivity. Waterside Gardens incorporates two highly productive models of small-scale agriculture that have proved themselves to be effective historically in peri-urban agriculture: chinampa-style canal-side gardens (Mexico city)[33] and the French intensive market garden (Paris).[34]

- In the gardens that use the inlet directly, hydrologically controlled subcanals between garden beds divert water from the adjacent inlet canal. These alternating strips of water and land crops are managed to make the system highly productive in several ways:
- Constant sub-irrigation of the growing beds;
- Aquaculture production from a self-feeding, integrated system of water plants and animals;
- Surplus fertility from the aquatic system in the form of muck dredged periodically from the canals for the adjacent bed soils;
- Temperature stabilization from the waterways that improves daily crop growth and extends the growing season.

Figure 17. Waterside Gardens (artist's conception by Jane North)

Farther from the water lie the frame and cloche beds characteristic of the French intensive method. Despite the development of biomass-based plastics, competition from higher priority biomass uses like food and heat has prompted a return to the French tradition of glass for frame covers and the bell-shaped cloches that create the microclimates to protect beds and individual plants.

Windmills pump canal water into raised tanks to provide a constant reserve of gravity-fed irrigation water. Adjacent ponds capture and biocleanse storm water that runs off the city's hills, constituting a water reserve that makes the system resilient to drought.

Another input essential to the intensive method is a constant and copious supply of fresh manure that is placed under and around frames and cloches to maintain growing temperatures in these all-season gardens. Initially the only manure source was the small population of livestock that peri-urban production systems can integrate. However, diminishing supplies of fossil fuel and limited supplies of local fuels like biogas from municipal black water processing have driven local transportation partially to rely on animal power. A growing mule population now transports people and produce around the county, much of it efficiently on the rebuilt light rail network. Like other peri-urban farms, this one provides stables for some of the mule contingent in return for the steady supply of hot manure. Their hay is transported by water directly into docks at the garden site from farms around the lake.

Wind protection is part of the intensive gardening system. The old supermarket and the high hedges on the

33 http://en.wikipedia.org/wiki/Chinampa
34 John Weathers, *French Market Gardening* (1909). http://ia331426.us.archive.org/3/items/frenchmarketgard00weatrich/frenchmarketgard00weatrich.pdf

northeast and northwest edges stop the coldest prevailing winds, and low walls throughout the gardens reduce wind at plant level while letting in sun.

While much of the French system is possible in urban agriculture, peri-urban spaces allow its full development as it originally functioned on the outskirts of Paris. This is because its year-round production requires quantities of hot manure as well as the constant attention of full-time gardeners highly skilled in the careful timing of watering, frame and cloche ventilation, and protection of frames from sun and cold. This garden recaptures the full knowledge- and management-intensive qualities that made the Paris market garden system so successful.

A more extensive system. The co-op includes a third, more extensive gardening system to grow crops like roots and tubers that need more space and to integrate small animal production. To fertilize this garden, the co-op manages a facility in which pig turners enhance the vermicomposting of part of the city's segregated organic waste stream.

Originally judged a brownfield, the soil of this part of the market garden spent its first years of conversion to agricultural quality under intensive grazing alternated with heavy applications of compost seeded with fast growing forages in the cleansing process described earlier. Now it consists of beds long enough to be worked by some of the mules housed in the co-op and grassed alleys wide enough to permit farm vehicles and grazing with rabbits and poultry in movable pens, as illustrated in Figure 18. In season, the rabbits thrive on an all-grass diet, and feed for the poultry is supplemented with part of the garbage and worms from compost production. The alleys are lined with composting sheds to which the poultry have access as their grazing pens are moved along the alleys. In all seasons the pigs, poultry, and rabbits consume the co-op's garden waste as one of their roles in the system.

Figure 18. Grass-fed rabbit production at Northland Sheep Dairy, a farm near Tompkins County

The old supermarket now serves many new functions. In addition to the stables, it houses farm tools and machines and harvest and feed storage areas. It also includes community centers to market products from adjacent community gardens, train new farmers, and house full-time farm workers and food processing centers. The south front is a passive solar greenhouse that heats the building and grows vegetable and nursery transplants for the rest of the farm.

Boundaries of the tripartite farm as well as individual beds are specifically designed for multiple functions. They include habitats that attract beneficials and trap pests before they reach food plants; bird and bat houses; flowering plants to attract pollinators; food bearing bushes, trees, and trellises that act as shelter belts against wind and sun; and walkways and benches to function as a parkland that brings urban residents into contact with the gardens.

As with much of peri-urban agriculture, the size of this co-op creates heavier seasonal labor needs than city gardens. With a large city population close at hand, however, it manages to attract enough seasonal workers by paying them with credits they can use when they purchase the food products of the enterprise.

RURAL AGRICULTURE

The last of the three overlapping food production systems that must be integrated to provide for the food needs of our county is rural agriculture itself, organized into a system of feeder farms associated with satellite hamlets in the hinterland of the city. In what follows, I will consider the needs and resources that will shape the design of future agrarian communities sharing a symbiotic relationship with the city of Ithaca and will offer a case study as a design example.

A general agricultural model

In rural parts of the county, space and other resources provide the opportunity to redesign agriculture most fully according to the general integrated system model described above. Moreover, the many existing or reclaimable wetlands in the county offer the prospect of sustainable systems on the model exemplified by the colonial farming system of Concord, Massachusetts referred to earlier. In colonial times, many agrarian communities in the Northeast made this grassland form of chinampa-style agriculture the core of their farming system. Communally managed wetlands were central because they *sustainably* produced the fertility that drove the system, indirectly via hay and thence manure, and directly from muck dredged from the canals:

These wetlands required considerable hydrological manipulation to make them productive, and they were transformed to a carefully managed resource in many towns. Extensive systems of drainage ditches, sometimes connecting for miles, rendered the meadows firm and accessible for teams during the mowing season, whereas dams, dikes, and road causeways provided hydrological control and augmented fertilization from natural flooding. Mowing, burning, and grazing, in combination with manipulation of the water table, shifted the composition of many wetlands from tree and shrub dominated to a cover of desirable grasses and sedges. The meadows returned a reliable yield of rather coarse hay, along with a rich muck that was cleaned from the ditches in the fall, dried, and carted to the barnyard or plow land.[35]

In land systems both wet and dry, grazing species such as the multi-functional cow formed the core of agriculture in colonial New England and sustainable agroecosystems in Cuba and elsewhere. They will likely be central to rural farming systems designed to survive the petroleum era.

A reconfigured social topography

Changes in rural land use, while not directly the subject of this essay, should be considered when envisioning a new plan for agriculture. If, like earlier societies that lacked fossil fuels, our society must use less energy to feed more people, it will require smaller, denser population centers with residences close to places of work. This constraint applies not only to cities such as Ithaca, but also to peripheral feeder towns and to the social topography of rural agriculture. In the US, cheap energy, cheap land, and the individualist ethic of "every man his castle" modeled on the European ideal of a landed aristocracy spawned a pattern of suburban sprawl on one hand and isolated farms on the other. In recent decades, the farms had to grow larger and even more isolated to survive in an agricultural economy where agribusiness multinationals exert monopoly control.

The traditional pattern in Europe is markedly different: apart from estates left over from feudalism, rural populations in Europe are even now clustered in agricultural towns and villages that include the farm residences and barns of many of the farmers who go out to work the surrounding land.

Energy descent planners in the US, including ecovillage advocates like Ithacan Robert Morache,[36] have made a strong case for converting to the European model of rural population centers, because, unlike suburban sprawl, this model clusters both farm and non-farm rural populations to make efficient use of energy, land, and

transportation resources that link to nearby urban centers. Ideally, these farming villages, circled by their farmlands, will replace present configurations of land use, in particular suburbia and many of the remote farms operated on the industrial model, both of which are unlikely to survive the end of the oil era. Whether our society will have the material resources or the political will to make such a complete conversion is an open question at this point. See the TCLocal article "Post-Peak Land Use Part 2: The Country"[37] for more detail on the farming village model.

Visioning a satellite farming village case: Lansing Landing

Imagine a once-thriving farming village connected to the county seat by good water, rail, and road transport routes that had in later times become a bedroom community. Now revived as a satellite ecovillage, buildings that serve a variety of agricultural, residential, and service functions are densely clustered in a hub surrounded by land devoted to diverse but related farming enterprises. Individual families and private cooperatives manage the enterprises within the general goals and guidelines set by the community and the county. Along with the community's commercial agricultural output, many households are engaged in homesteading production from kitchen gardens and small-scale animal husbandry. The village is planned with a systems design, well illustrated in the permaculture movement, which uses both food and nonfood species for the greater health of the farming community and its ecosystem: it organizes them functionally, spatially, and temporally in a calendar with a decades-long time horizon to serve this goal.

Today's ecovillages have made a start on the agro-integrated design that will be required here in the future. Figure 19, based on a study of the Ithaca Ecovillage,

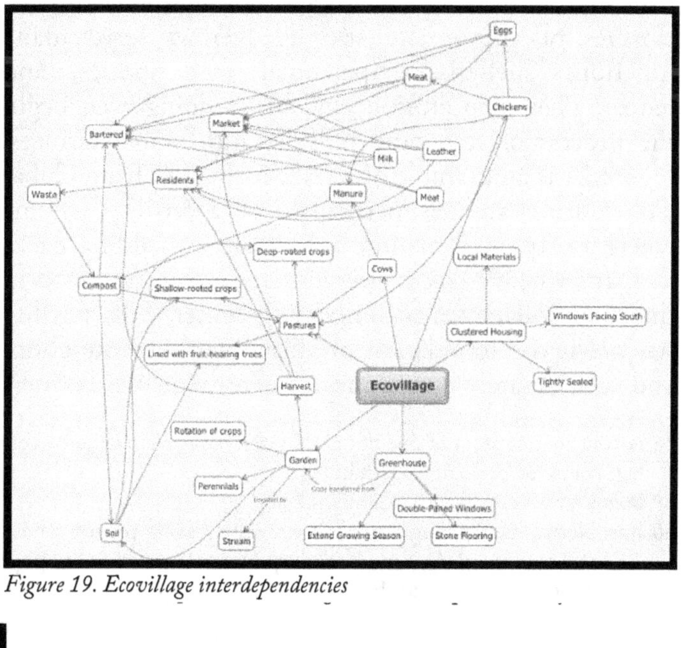

Figure 19. Ecovillage interdependencies

---

35 Charles L. Redman and David R. Foster, *Agrarian Landscapes in Transition: Comparison of Long-Term Ecological and Cultural Change* (Oxford: Oxford University Press, 2008).

36 Morache's plan of village clusters in the urban hinterland includes farms, residences for urban workers, and enough commerce to support a population of 450 households. www.chrysalisconcordium.org

demonstrates some of the flows, interdependencies, and synergies that can be captured in a farming ecovillage designed as an integrated system.[38]

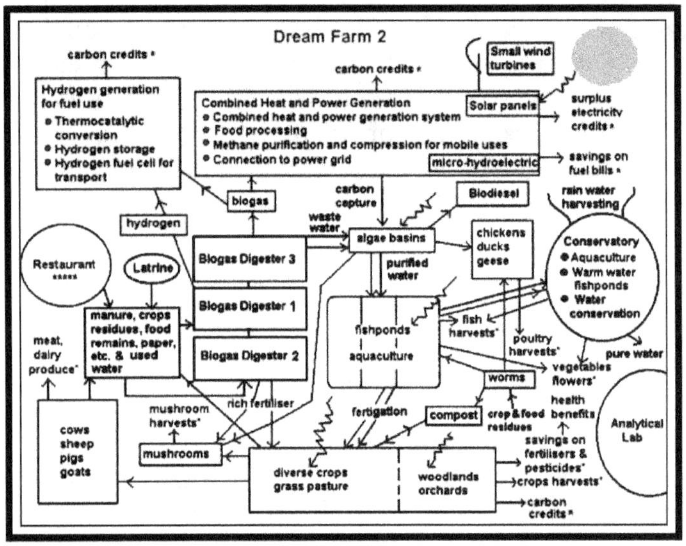

Figure 20. Functional integration in a planned agricultural community

Lansing Landing builds on the example of many ecovillages today, but aims for a higher standard of sustainability, including the need for greater heat and energy self-sufficiency; affordability (many ecovillage dwellings are too expensive for the average person); diversity of functions, including farming as the core function; and more complete recycling (how many ecovillages collect and process night soil?). Some of the components and functions present in the community envisioned here attain the high level of integration planned for an agricultural community in the United Kingdom by the Institute for Science in Society, as illustrated in Figure 20.[39]

Fertility. Open, sloping land plays an important role in the village agroecosystem. As described earlier, animals graze a hillside system[40] of perennial forages dotted with food-producing trees. Hedgerows crisscross this landscape, surrounding fields and carving them into enclosures of appropriate size. Hedgerows serve many functions: shelterbelts, perennial food species, and fences. They stop erosion, and by so doing even begin the process of reshaping hillsides into arable terraces. Figure 21 is an example of terrace formation from Cuba. The grazing animals participate in a fertility scheme where a surplus of manure is built up as bedding packs in barns where stock is overwintered, then processed in the main village vermicomposting center. This fertility scheme is the foundation of village wealth production, and so ultimately determines its quality of life. Farmers

also use the grazing animals to optimize biomass production in row crop acreage whenever the acreage is in a grass rotation.

Along with biodigested humanure from the village and the city of Ithaca, applications of compost made from winter livestock manure and bedding create the tight nutrient cycling that builds and sustains the fertility of the land. Manure and village sewage that is more conveniently handled as liquid is fed through a fuel-producing biodigester, then solids separators followed by cleansing ponds that grow duckweed for high protein animal feed, and finally back to fields as in Figure 20. Village farmers use a sophisticated scheme of fallows, rotations, and winter- and roller-killed cover crops to further control fertility and weeds with minimal tillage.[41]

Water and wood. In Lansing Landing, ponds have been placed high on the hillsides to capture spring water and runoff for many uses: village and livestock supply, water power, and irrigation, to name a few. Lower ponds recapture water for additional uses: recreation, fire protection, and a village reserve. They function as part of a water management array of berms or swales, like the keyline plan described earlier, that keep water working within the watershed as long as possible.

Figure 21. A hedgerow in Cuba stopping soil movement on a slope

Drawdown of forest resources to the point of crisis occurred repeatedly in European and U.S. history before the oil age, when biomass was the main source of energy. Forest cover in Tompkins County dropped from almost 100% in 1790 to 19% by 1900, then increased to 28% by 1938 and to over 50% in 1980.[42] Most of the loss of forest cover can be attributed to a combination of logging for firewood and timber and clearing for livestock

---

38 A contribution from of one of my students, Jason Fleischer, in a college course on ecological agriculture.

39 http://www.i-sis.org.uk/DreamFarm2.php

40 Karl North, "Optimizing nutrient cycles with trees in pasture fields," *LEISA Magazine,* 24/2, June 2008. http://www.leisa.info/index.php?url=getblob.php&o_id=209102&a_id=211&a_seq=0

41 Pioneered by Pennsylvania vegetable farmers Anne and Eric Nordell and archived in their ongoing column, "Cultivating Questions," that dates from the 1990s in *The Small Farmers Journal,* Sisters, Oregon.

42 Bryce E. Smith, P. L. Marks, and Sana Gardescu, "Two Hundred Years of Forest Cover Changes in Tompkins County, New York," *Bulletin of the Torrey Botanical Club,* Vol. 120, No. 3 (Jul. - Sep., 1993), pp. 229-247.

production and other agriculture. The much bigger present county population will make far greater demands on forest resources. It would be mistaken, therefore, to assume on the basis of current forest cover that the county can rely on wood for its future energy needs.

The village actively manages enough forestland to do its part in providing county forest product needs, among which firewood for heat and timber for shelter are paramount. By replacing the extremes of no management and monoculture that were luxuries typical of an earlier era, active management stimulates both biodiversity and production in a balance to achieve a wide range of agroforestry goals. Many forests are maintained on ridge tops and uplands for the health of the watershed. Groves near the village center create useful microclimates, temper prevailing winds, and provide for recreation.

Food and Fiber. The imperative of energy efficiency has gradually reconfigured land use in this village to cluster the more intensive agricultural activities in the flat, most fertile land ringing the village center. This circle contains the rotating fields of starch staples, vegetable polycultures, meadows for the most intensive animal husbandry, and fibers like hemp and flax. Its output of foods and fibers that traditionally grow well in the region help ensure the food security of the county.
Crops like flax and hemp, which produce fiber, oil, and other ingredients of manufactured products such as paper, clothing, paints, and preservatives have reappeared as competing petroleum products have disappeared and competition for forest products has increased. Different parts of the hemp plant produce flour and oil for food, paper, and composites, including boards that reduce logging pressure on forests, rope and cloth, lubricants and other petrochemical substitutes, and important nontoxic medicines. Hemp productivity per acre is four times that of sustainably harvested wood, and twice that of cotton-without cotton's need for pesticides.[43]

Not far from the village is a wetland modified with canals and ponds to grow aquaculture crops. Because of the constant source of crop water, the wetland system is an anchor that guarantees a reliable source of forage and bedding for livestock both in the village and in the peri-urban animal enterprises.

Part of the wetland has been developed into a true chinampa-style production system. As described earlier, the chinampa configuration of aquaculture in canals surrounding raised fields is integrated in a way that ensures higher productivity over dry-land agriculture. While most examples of this system come from Central America and Southeast Asia, the system has also succeeded in northern Japan in a water-moderated climate similar to

ours in Lansing Landing. Figures 22 and 23 from Japan demonstrate some of the possibilities.[44]

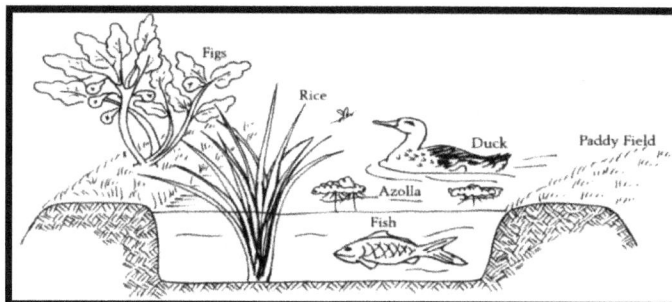

Figure 22. A rice-fish-duck-azolla system. Azolla (duckweed) is a floating fern that fixes nitrogen and produces protein

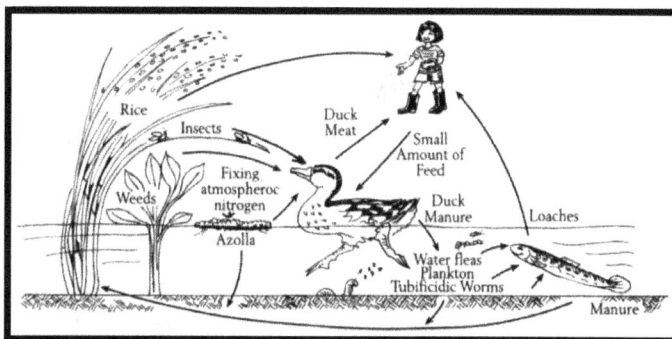

Figure 23. Material cycles of azolla + loaches + ducks + rice. The system produces rice, duck meat, duck eggs, and fish for a small input of feed

The core of village livestock husbandry at Lansing Landing is the dairy enterprise, much of which has returned to the energy-efficient model of seasonal, grass-fed milk production from the hillside pastures and hay fields. Breeds chosen to fit the system are hardy, dual-purpose, and smaller than the energy-intensive breeds of the industrial agriculture era that were designed to maximize production at any cost. Cows, sheep, and goats are pastured along with work mules and horses in a multi-species grazing system that benefits from the complementary grazing functions of the different species. Dairy and crop byproducts sustain some pig and chicken production. The level of animal production is determined by the role of animals in supplying ecological services to the community's agriculture, not by county demand for animal food products, which is currently excessive and unhealthy. At Lansing Landing, the level of production of animal foods is closer to what is needed for a healthy human diet.

Like animal genetics, the genetics of the crops grown by the village have changed to reflect the exigencies of the post-petroleum era. Instead of hybrids that sacrifice local seed control and the resilience that a large gene pool provides, village farmers, employing traditional selection methods, have developed open-pollinated seeds that they can save and share. While yields from savable

43 The 1995 documentary film *Hemp Revolution*. Anthony Clarke, director.

44 Takao Furuno, *The Power of Duck* (Tasmania: Tagari Publications, 2001).

seeds can rival the productivity of hybrids,[45] village farmers have selected for both plant and animal types that balance productivity with traits like hardiness and other low-maintenance characteristics.

Village Enterprises. Even closer to the center, to be within walking distance of their workers, are animal and crop barns, village-scale composting and biogas digester sites, tool manufacture and repair shops, and other agricultural support facilities. One example is a piggery used to turn compost. Fed largely from dairy byproducts and kitchen garbage, its manure in turn feeds a small biogas generator like the one in Figure 24.

Figure 24. Biodigester made with one layer of plastic tubing 1.2 m in diameter and 6 m long, connected to a pig pen with 20 animals and fenced with Mulberry tree. Finca Ecológica Tosoly, UTA Foundation, Guapotá, Santander, Colombia. Photo: Lylian Rodriguez

Processing plants that preserve raw farm products while reducing water content to make them more transportable are village enterprises that serve an important function in the county food system. Examples include the conversion of milk and fruit to aged cheese and preserves and the lumber-drying sheds at sawmills. Near the center of town is the village recreational fish and skating pond, one of the ways a stream running through the valley has been harnessed.

One of the important functions of the village is to recruit and train new farmers from the urban population to run the more labor-intensive agriculture of the new era. An educational complex serves as a public school for the village, an agricultural research and farmer training center, a farm camp for urban youth, and an adult farm camp for harvest volunteers and vacationers from Ithaca. In turn, the village draws on urban populations for short pulses in labor needs, like haying and other harvest activities that must be accomplished in a brief window of opportunity.

Rural agriculture and the county food supply

This series has described three types of area agriculture needed to sustain a county population of 100,000: urban, peri-urban, and rural. Of these, rural agricultural systems will be of primary importance. Urban and peri-urban gardens can provide quantities of fresh vegetables and fruits, but only rural farms have the space to grow enough of the starchy staples like potatoes, grains, beans, and rice that have historically supported urban population densities. Moreover, only rural farms can supply enough of the materials like oils, fibers, and wood that are basic necessities in our cold climate. Agrarian villages, not the urban center, will again become the heart of a relocalized county food system in the coming years.

C

Chickens in the Energy Descent

By Tom Shelley (March 2011)

INTRODUCTION

Birds and their eggs have been part of our food chain for tens of thousands of years. In hard times, birds and their eggs were survival foods. In the not too distant future, chickens will be a pillar of survival and resiliency as we proceed into what we believe to be a looming energy descent. Chickens are comparatively easy to raise, and they provide high quality meat and eggs all year round. Some writers prefer ducks,[1] and ducks are an important contributor to the small farm environment, but well-managed chickens are a better fit as an integrated component of a sustainable farming system.[2]

Raising chickens for eggs provides a highly versatile source of protein. Eggs can be stored for a reasonable period of time with relatively little energy input. They may be sold or traded for other goods. In her new book, *The Resilient Gardener,* Carol Deppe defines five crops you need to "survive and thrive—potatoes, corn, beans, squash and eggs." This article will consider some of the parameters for raising chickens, explain how these parameters will be affected by the energy descent, explore some alternatives for current practices, and offer many questions still to be answered.

WHICH CAME FIRST, THE CHICKEN OR THE EGG?

In the energy descent, raising chickens first for eggs and secondarily for meat will be a preferred strategy, for three reasons. First, the nutrients in eggs are denser and more complete than the bird's meat itself; second, eggs can be stored or preserved fairly easily for future use; and third, eggs have more versatility for food prepara-

tion than just the chicken meat itself. Mayonnaise, custards, etc., depend upon the chemistry of the egg to make a unique food product. Roosters, roughly fifty percent of hatchlings, are generally reserved for meat birds, as are hens that are no longer productive.

There are currently 113 breeds of chickens recognized by the American Poultry Association.[3] Many more varieties and strains of chickens are available, and the selection of the appropriate chickens can be a daunting task. Laying hens are selected for their egglaying productivity, length of their productive years, heartiness, body size, egg color, temperament, and other factors. I am familiar with Black Australorps, but Plymouth Rocks, Orpingtons, Rhode Island Reds and a number of other breeds are reliable egg producers. Some aids are available for novice chicken owners to help with breed selection.[4]

Figure 1. Black Australorps feeding

One distinct advantage of chickens (and many other fowls) is their ability to easily breed and brood eggs to make more laying hens. There are many devices that have been invented to incubate eggs to produce chicks. In the energy descent, especially during times of crisis, it may not be possible to incubate eggs with electrically powered equipment. Since it is fairly easy for chickens to produce and raise their own young, it is strongly advised that all small scale chicken raisers learn how to breed and brood their own chicks to ensure a sustainable supply of laying hens. Several good references for raising chickens from eggs have been published. One highly recommended book is Gail Damerow's *Storey's Guide to Raising Chickens.*[5]

---

1 Carol Deppe, *The Resilient Gardener* (Chelsea Green Publishing, 2010), p. 178.

2 See the TCLocal series of articles on local food production by Karl North, beginning with "Visioning County Food Production—Part One: Introduction" (http://tclocal.org/2009/07/visioning_county_food_producti.html).

3 See, for example, http://139.78.104.1/breeds/poultry/, from Oklahoma State University.

4 See, for example, http://www.mypetchicken.com/chicken-breeds/which-breed-is-right-for-me.aspx

5 Storey Publishing, 1995. For additional general information see http://www.backyardchickens.com/lcenter.html and http://www.lionsgrip.com/pastured.html; *Small-scale Poultry Keeping* by Ray Feltwell (Faber and Faber, 1992); and the periodical *Backyard*

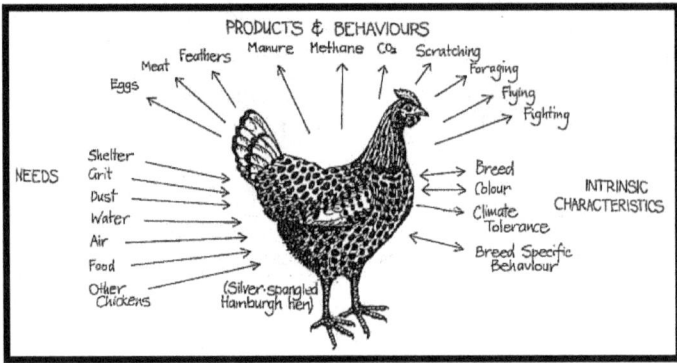

Figure 2. Selecting a breed (Mollison)

Day-old chicks are widely available from many sources, local, regional, and national. They are going to be increasingly expensive in the future due to rising transportation costs and newer food security regulations. In a steep energy decline, the traditional regional and national sources may no longer be affordable or even available at all. We will need to depend upon home-brooding or smaller scale, local commercial brooding/incubation of chicks.

HOUSING OPTIONS

Many beginning chicken owners have romantic notions of "free-range" chickens. Free-range chickens usually have at least a 50 percent loss rate due to predators. Also, the eggs of free-range chickens can be difficult to gather because they are often laid in hidden, inaccessible places, greatly reducing the useful yield of the flock. If you are dependent upon chickens and their eggs for a subsistence food base, "free-range" is not a good idea.

To maintain high levels of productivity and prevent predation, chickens must be watched over and, at a minimum, contained within a secure fence high enough to keep the chickens in the pen. An electrified fence powered by a small solar panel will prevent almost all small-animal predation.[6] These portable fencing systems, while not inexpensive, allow for frequent reloca-

Figure 3. Chickens behind electro-net fencing

tion of the fence to enable appropriate management of the areas being pastured. Netting the penned-in area may be needed to prevent predation from hawks. Chickens kept in fenced-in enclosures are said to be pasturing or "free-roaming," but they are not free-ranging.

The construction and use of a secure chicken coop that is small-mammal proof is strongly recommended. Our chickens roost in their coop and are tightly closed in at night to prevent loss from predation and to provide shelter during bad weather, especially over the winter. Chicken coops can be made from a wide variety of materials, from hundreds of available designs.[7]

Figure 4. Chicken coop on skids

The use of chicken tractors is also very popular and greatly extends the functionality of having chickens while providing additional protection and security. A chicken tractor is a lightweight, movable coop. Many designs are available, depending upon your use of the tractor and the number of chickens involved. *Chicken Tractor,* by Andy Lee and Pat Foreman,[8] gives extensive information on the construction and use of chicken tractors. Chicken tractors can be used to pasture chickens, with the tractor being moved to fresh grass as needed. If tightly constructed and installed, a chicken tractor will provide reasonable security against attacks by small predators. Our chicken tractors are a wood frame covered with chicken wire with a hinged piece of plastic sheet roofing material for a lid. The lid is normally hooked shut when the tractor is in use. Some people use small hoop houses for chicken tractors.[9]

The best use of the chicken tractor is to prepare an existing garden for planting. Six to ten chickens in one of our chicken tractors will eat everything organic down

7 For examples, see http://www.freewoodworkingplan.com/index.php?cat=212. Sometimes the home page works better: http://freewoodworkingplan.com/

8 Good Earth Publications, 2006.

9 You can see a wide variety of chicken tractors at http://home.centurytel.net/thecitychicken/tractors.html

to the ground over a 4 x 8 foot area in 10 to 14 days, including all sorts of difficult-to-eradicate weeds and grasses. We then loosen up the area with a fork to remove the big roots of last years' crops and weeds while mixing in the chicken manure and some additional compost. Other techniques are possible, such as working up a new garden plot or feeding specific home-grown crops to chickens.

Figure 5. Chicken tractors in the garden

FOOD AND WATER

For many small-flock chicken owners, the cost of traditional grain- and soy-based feeds is 70 percent of the cost of the maintenance of their flock.[10] Commercial grain products used in chicken feeds consume vast amounts of fossil fuels in their production, processing, and distribution. Overall, agriculture contributes eight percent of the anthropogenic component of global warming gases. Even a modest rate of energy decline will have disproportional impacts on the cost of laying mash, pushing the price of chicken feed out of the range of feasibility for many small flock owners. This is already happening. A steep rate of decline would mean that nicely milled and amended layer mash in 40-pound bags may no longer be available at all.

Fortunately, for those who can develop a flexible and resilient approach to feeding chickens, many options to currently available commercial chicken feed are available. Chickens will eat almost anything, with some major exceptions (alliums and citrus in particular). If chickens are free-roaming and pastured and given a variety of supplementary foods, they will eat those foods that provide adequate nutrition. Many small flock owners feed mostly kitchen food scraps and some scratch feed (cracked corn or corn/wheat mix) and have healthy chickens and lots of great eggs. Owners of larger flocks cannot supply enough scraps to provide adequate nutri-

tion, so they traditionally resort to commercially available feed mixes.

As the energy decline progresses, access to commercially available layer mash will be increasingly limited for the small flock owner due to increased costs, limited access to some ingredients commonly used in commercial feed mixes, and other factors (the difficulty manufacturers may have in repairing or replacing equipment, for example). If land is available, many crops can be grown for chicken feed with low technology and few investments. Since chickens will eat everything from amaranth to zucchini,[11] there are many options, depending on the type of soil and the availability of water and nutrients (compost), seeds, labor inputs, etc. Carefully selected crops, most of which are human food crops as well, will allow for adequate nutrition for a flock of chickens over the seasons. Larger flocks are going to require large plots of land, with more grains and seeds to be grown and saved for the winter.[12]

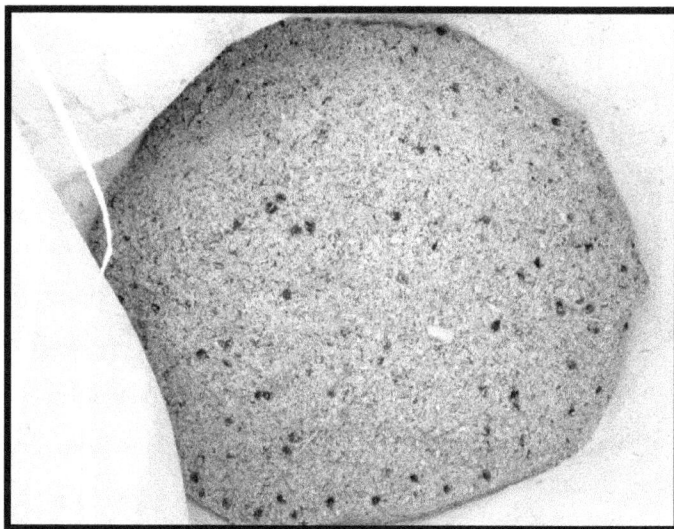

Figure 6. Locally made layer mash

Other local grain and feed options are readily available. For example, I have been purchasing "waste" grain products from Farmer Ground Flour in Trumansburg, New York. I mix supplements with the waste grain products and make a high-grade, organic layer mash. The carbon footprint of my homemade layer mash is significantly less than feed from other regional or national outlets. Other nearby grain mills have sold "seconds" or waste products to local farmers over the years for chickens and other farm animals. I anticipate that feed co-ops will develop to split up the rising costs of the components of feeds. Fish and crab meal, for ex-

---

10 *Storey's Guide to Raising Chickens*, p. 53.

11 For example, http://steephollowfarm.wordpress.com/2009/06/18/chickens-like-a-lot-of-things/. See "Local Notes on Chicken Feed" (http://tclocal.org/docs/chicken-feed.pdf) for some ideas about local possibilities.

12 See Gene Logsdon, *Small-Scale Grain Raising* (Chelsea Green Publishing, 2009).

ample, are commonly used feed amendments. Being in the interior of the country, traditional sources of fish meal would be either very expensive or non-existent, depending upon the slope of the energy decline curve. Perhaps a local source of farmed fish for fish meal could be developed.

Cooperative efforts to share resources for chicken feed would be very useful. Sharing bigger farming equipment, sharing saved seeds, and trading chicks to maintain diversity are examples. Chickens love milk, yogurt, and other dairy products. Apples, pumpkins, squash, and other fall veggies that store fairly well can be fed over the winter, providing diversity when pasture isn't available. Waste vegetables from nearby farm stands can often be gleaned in the summer and fall, and they add value to the nutrient intake of your flock. Chickens love hay in the winter, and we have local sources of organic alfalfa hay. Duckweed, which commonly grows on local ponds, is a highly nutritious chicken food (see the article "Envisioning Tompkins County Food Production" earlier in this volume for more about duckweed in sustainable food production). Sprouted grains provide grass in the winter; I use oat grass, because oats are very inexpensive and germinate readily.

Figure 7. Chickens eating oat grass

Other options include feeding chickens active compost or certain insect larvae. Active compost has a high percentage of insects and other high-protein sources perfect for chickens. Some chicken farmers raise meal worms or black soldier fly larvae as chicken feed.[13] These techniques can significantly reduce the consumption and dependence upon grain-based feeds and their high fossil fuel inputs and large carbon footprint.

Chickens need a lot of water. Laying hens use up to two cups of water per day and even more in hot weath-er.[14] Water requirements are often higher in winter, when humidity is low and feeds and grasses are dry (hay, alfalfa cubes). The chickens' water needs to be clean, potable water from a reliable source. Springs, wells, and urban water sources are all commonly used. In the early stages of energy decline, most of these sources will remain stable, although spare parts for pumps and wells may be hard to obtain at times. In a steep energy decline, energy sources and systems (delivery of public water supplies) will be disrupted or non-existent, parts will be impossible to obtain, and only secure natural sources (uncontaminated springs or wells) will allow for good quality water. Alternatives need to be developed. Clean rainwater catchment on a scale to water a modest flock is possible for most chicken owners.[15] The catchment and storage systems would need to be in place and functional when needed. Chickens can also drink from a clean stream or pond if one is available. Contamination from agricultural runoff, especially if you are raising organic chickens, and from animal wastes is of serious concern when using a stream or pond as a water source.

Other requirements for chickens are oyster shell and grit. Layer hens have a high calcium uptake, and the general recommendation, based on information from Lakeview Organic Grain, is 127 pounds of crushed oyster shell per ton of feed. The grit, needed to grind food internally, is most frequently sold as ground granite. Grit is free-fed; in a free-range or free-roaming situation most chickens will find all of the grit they need outside. Grit is most often fed in winter, when snow cover and frozen ground prevent normal foraging. Oyster shell will be more problematic, especially in a steep energy decline, and alternative materials and sources need to be found. Ground up egg shells provide one option, but this is a limited source.

FLOCK MANAGEMENT

Many management issues will be affected by energy descent. Moving a coop from pasture to pasture is easy if you have an appropriately sized tractor and the fuel to run the tractor. If not, do you have a neighbor who owns a horse who will help you every two weeks or so? Some may elect to have their own horse; perhaps several nearby farms could share a horse and the expense of maintaining the horse. Building a coop with wheels would facilitate movement, perhaps only requiring a few strong people.

---

13 See http://www.sialis.org/raisingmealworms.htm#timetable and http://blacksoldierflyblog.com/

14 *Storey's Guide to Raising Chickens*, p. 60.

15 See, for example, Bill Mollison, *Permaculture—A Designers' Manual*, 2nd ed. (Tagari Publications, 2004), pp. 165-170. A detailed overview of rainwater catchment techniques developed in third-world countries can be found in John Gould and Erik Nissen-Petersen, *Rainwater Catchment Systems for Domestic Supply* (ITDG Publishing, 1999).

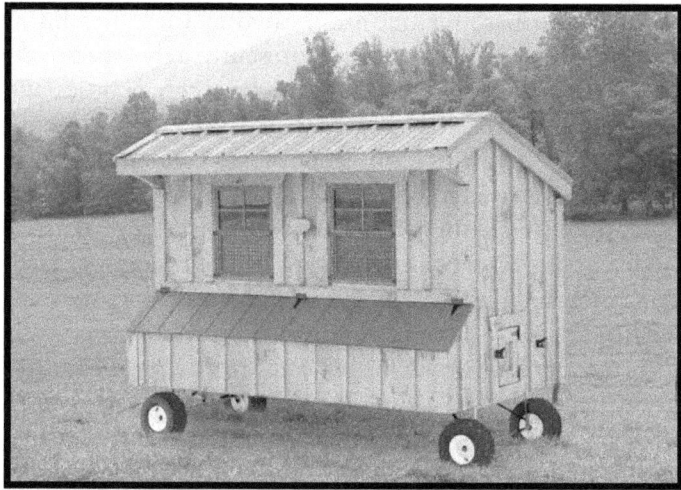

Figure 8. Coop on wheels

As the energy descent progresses and liquid fuels become more expensive, different business models will quickly evolve. We now deliver most of our eggs to our CSA customers. Perhaps there could be a central pickup point arranged for several customers in one area so as to save on fuel costs. Feed co-ops will develop to split up the rising costs of the components of feeds. Joint ownership of expensive equipment will enable a number of farms to thrive.

As eggs become an increasingly valuable component of our diets, food security issues become more important. There are well-established protocols for the handing of birds and eggs that reduce the possibility of contamination and disease transmission.[16] This is an area that is often poorly understood by small-flock owners.

Keeping chickens cool in summer and warm in winter is important. The use of solar heating and solar PV for other equipment will become increasingly im-portant. If global warming increases at the rates projected, it may be necessary to change the breeds or varieties of chickens and other fowl to those that tolerate heat better than traditional breeds.

Attention to manure management will be especially important. Chicken manure is very rich in nitrogen and other nutrients. It may be composted or mixed directly into the soil. The appropriate rotation of fertilized pastured areas, gardens, and grain plots can maximize the inputs of chickens to the nutrient cycle. In a steep energy descent, most or all of our resources will come from the farm itself.

The management of breeding stock, the culling of poor producers, and other hands-on management issues will need to be addressed to maximize the long-term success of the flock. Health issues will may be more of an issue, especially in a steep energy descent. Our chicks come from Missouri, arriving with vaccinations for several diseases. Where would these vaccinations come from in the future? Very few vets know about chickens. Last year New York State defunded the state veterinarian position that has served the poultry industry in New York for several decades, so there is only thin support of flock health available going into the near future. How are we going to learn to be our own vets as far as our flock health is concerned? There are many challenges ahead.

CONCLUSION

Readers of this article may not live to see the hard times ahead, but their grandchildren certainly will. Chickens and other fowl will be an integral component of a resilient community as we enter an uncertain future.

---

16 http://www.eggsafety.org/producers/food-safety-regulations

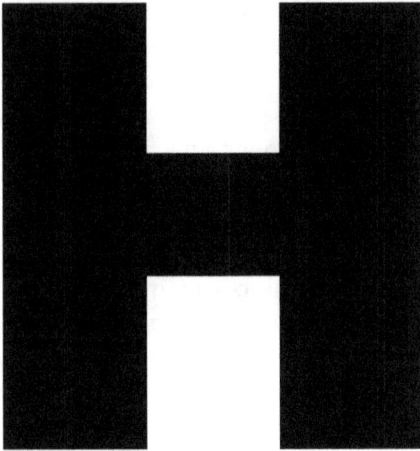

Health Care in an Energy-Constrained Environment

PART TWO: OPTIONS FOR RE-EVALUATING CARE RESOURCES

By Bethany Schroeder (November 2009)

In Part One of this two-part series on healthcare resources in Tompkins County (available online at tclocal.org and reprinted in *Thinking Local in Tompkins County, Vol. 1*), I pointed out that today we have a variety of options and a well-developed infrastructure to meet the health needs of many local people. Noted exceptions include un- and underinsured residents, now estimated at 13,000. Some of these people are treated outside the County at regional medical centers and some receive care at the Ithaca Free Clinic (IFC). Many do without regular care at all, visiting a facility on an emergency basis only. As an unstable economy and reduced resources persist and worsen, more and more people will experience the exigencies of decline. In terms of health care, much can be done to mitigate the effects through evolved expectations and planning for the change. Recognizing health care as a right rather than a privilege goes some distance toward effective planning. Understanding that the illness or injury of anyone in our community is a hazard to all of us and one we should address by providing support also demonstrates our humanity and our solidarity.

OVERCOMING BARRIERS TO CARE

In the future, barriers to care will include transportation challenges, lack of available facilities, and alterations in care models.

Transportation to care

Whether one is a healthcare worker or resident in need of services, a chief barrier to care delivery in the future will be transportation. Because most large health resources such as Cayuga Medical Center (CMC) and the Convenient Care Center are located at the outer limits of the city of Ithaca, people presently rely on private cars, taxis, bicycles, or public vehicles, such as buses and Gadabout shuttles, to get to and from appointments. CMC also operates a Convenient Care Center located in the adjacent county of Cortland, where some residents of Tompkins County living nearer to Cortland than to Ithaca access care. Similarly, most of the residential facilities, including assisted living homes and skilled nursing facilities (SNF), have private shuttle services specifically for transporting residents to and from healthcare appointments. As clinic and office spaces are developed in Ithaca, more and more physicians are presently located at the periphery of the city's boundaries. Nonetheless, already established physicians and nurse practitioners in smaller offices, as well as the offices of many complementary and alternative providers, are located within the city and within some of the villages and towns, making it possible for residents who live nearby to walk to appointments. Both options of locale have pros and cons, and these will change over time.

An obvious way of managing challenges to transportation, both from the perspective of caregivers and people needing care, is to encourage healthcare workers to live near worksites and for residents to establish relationships with providers near their own homes or worksites. The present centralization of healthcare facilities makes this difficult to achieve, whereas planning for a future change now could make the concept more acceptable. Specific transportation options are outside the scope of this article and will be addressed by other TCLocal contributors. Nonetheless, an obvious consideration includes developing employment and social structures that routinely allow workers to seek care during work hours, especially important to workers in settings located near care settings. At the same time, healthcare providers could consider holding flexible hours in order to facilitate access through available transportation options.

Another option for arranging transportation is to reverse the process, especially in clusters of dense dwellings. Teams of caregivers in any number of configurations could easily walk through neighborhoods delivering service—either in the form of direct care or education or both. Physicians have largely discontinued making house calls in the U.S., but visiting nurses still do travel to homes, and this practice may prove efficient in some circumstances and settings. For example, a team composed of a registered nurse, a dietician, and an herbalist could offer nutritional and medicinal education. A chiropractor, an acupuncturist, and a massage therapist could provide alternative pain management. If the teams worked together, they could help one another in the process of finding the right method of fulfilling the needs presented to them.

Tompkins County could also learn a lesson from the Cubans, who assign physicians, nurses, and others to live and work in specific neighborhoods, inspiring, according to reports, a deep commitment to the neighborhood and its residents. Care providers in the immediate vicinity of those needing care are naturally able to see and to know their prospective patients in a different way than when both reside apart.

If things get as bad as some of us think they might, another potential consideration is the option of taking care into the community, such as the former rural district nursing practice. At the beginning of the last century, nurses cared for patients in rural settings using horses to get to and from settlements. In many parts of the U.S., this would be an untenable scenario, whereas Tompkins County—indeed, the entire Finger Lakes region—already supports many horses, horse farms, and local routines that include horses in daily life. Under circumstances of energy descent, many more people may be occupied in agricultural pursuits, in which case we might expect more farming injuries and other agriculturally-related healthcare needs. Visiting nurses or even visiting physicians could well be a necessary part of daily life.

Care facilities

Part One of this series provided an overview of care facilities in Tompkins County. Apart from the Public Health offices, owned and operated by local government, most local facilities are privately owned. City and town planning boards review and approve the construction or re-fabrication of care facilities, and some degree of oversight of the development of facilities occurs through the work of the Health Planning Council and its advisory board. Projects that may rely on public money, such as Medicaid dollars used to house residents in assisted living homes or SNFs, are scrutinized for the need of services in a particular area. Nonetheless, there is no master plan based on realistic census projections and estimates of available resources necessary to ensure care for all residents.

In an era of adaptation, the leadership of Tompkins County can rethink the requirements of a care facility, as well as the number of facilities in any part of the county. If care can be delivered in less formalized and standardized settings, then almost any storefront or main floor of a house or other common building is adequate so long as it has bathrooms and a hand-washing sink in a common space, as well as space for reception and discharge activities, a waiting room, and a private room where primary providers can interview, examine, and treat people.

In Alexander's 1977 *Pattern Language,* the architect and writer advises: "Gradually develop a network of small health centers, perhaps one per community of 7000, across the city; each equipped to treat everyday disease." Identifying small or modest buildings or parts of buildings with multi-use features, such as several doors for entrance and exit, ground floor access, and a variety of plumbing options, could help to realize the image of "a network of small health centers."

Before the advent of cheap oil, providers living in neighborhoods delivered care in their homes, and the very sick or those who could not be transported received house calls from physicians and nurses. Hospital care was reserved for the gravely ill and was often an option of last resort, because families were loath to be separated from one another and hospital care was for many people prohibitively expensive. Organizing care within a matrix of walkable locations and within easy distance of one's home or work may even have the potential for making the idea of care less forbidding. Reserving the hospital for the most extensive and demanding care and, once energy descent is fully and inexorably underway, possibly reshaping the hospital for a variety of community roles, may be the most responsible use of resources.

Alexander has also suggested organizing health centers with recreational and educational activities related to good health in mind. Some of our local resources have exactly this level of functionality. Island Fitness, owned in part by CMC, includes fitness training equipment, offers a broad range of fitness and stress reduction exercise classes, and operates a spa with massage services, all the while providing physical therapy and rehabilitation to people who are strong enough to use an out-patient facility. Similarly, the Integrative Medicine offices in downtown Ithaca are within easy walking distance of the City Health Club, and a number of chiropractic offices in downtown Ithaca are located near pilates and yoga studios. Viewing these opportunities as part of our local resource and planning in a way that supports groups of services in clustered arrangements is good for the people who need the help and for the people giving it.

Care models

Most employers either provide or require a certain amount of on-the-job or continuing educational effort so that the knowledge within the workforce remains current. At this time, healthcare coverage in work settings of a certain size is mandated, and some progressive employers understand and appreciate that employees knowledgeable about matters of health and wellness have made an investment in their own longevity by demonstrating responsibility for their choices. By the same token, most schools offer classes in healthy living, sports and exercise, nutrition, and lifestyle. The person who has learned about his or her health needs and is willing to take steps to maintain a healthy status is an asset to the workplace, to the school, and to the community. Such an individual is also an example of the lessons of prevention taken seriously.

Many people already know much about their bodies. A by-product of our modern lives and the leisure we have includes aspects of self awareness that can lead to healthier states of being. Yoga, t'ai chi, qui gong, and many other martial and meditative arts support health and healthy living. Similarly, recreational and competitive sports have the potential for promoting self-discipline and long-term vigor. Prevention will necessarily be a big feature of healthcare delivery in a post-peak environment. The residents of Tompkins County are already better prepared than many people in the U.S. for the choices related to prevention: primary care, complementary and alternative medicine, regular exercise, sound nutrition, and a holistic perspective on the relationship between the mind and the body inform the lives of local residents.

On the other hand, most of the treatments, therapies, and surgeries we presently rely on as interventions to maintain or improve health require products made largely from petroleum. Under our current system, we take for granted the disposal of used equipment, if only because it's impossible to thoroughly sanitize or sterilize plastic containers and fixtures. In times past, most of the implements of care were made of glass and metal and could be refurbished and reused. Preparing to live with fewer of these adjuncts requires that we re-think our throw-away healthcare culture and take better care of the health we have now.

Much as energy descent will change aspects of care delivery, we can expect climate change to influence the illnesses we are exposed to. For example, as temperatures increase in presently cold climates, microbes and vectors that were previously unable to survive lower temperatures will begin to survive and then thrive. Treating diseases with which we have no experience and no immunities will require flexible and creative approaches, good diagnostic abilities, and an educated response not only from caregivers but also from community members. As is true with many of the illnesses we now confront, new illnesses from other environments often diminish in the face of prevention. In addition, we will need to learn to use netting to protect sleeping and resting spaces, effectively manage snakes and other animal interlopers, and contend with the effects of poisonous or otherwise noxious plants and insects. We can expect a benefit from such accommodations to be the return of better and more regularly used porches protected, of course, against the predations of new pests of one sort or another.

Two specialties in health care are especially well-suited to the delivery of services in a post-peak environment in which unknown illnesses and strained resources prevail. Emergency medical administrators and providers as well as public health officials and providers will be in much demand as energy descent and climate change reshape our world. Emergency medical professionals are already accustomed to the concepts of triage and developing priorities required to confront disasters and the shortages disasters incur. Public health professionals are also continually advised about the changing landscape within the regions that shelter their communities. Both specialties promote interdisciplinary models of care and encourage broad areas of expertise, and both could be called on to organize local efforts to safeguard populations and teach individuals how to respond to the threat of disease. These professionals invariably know how to think about dealing with shortages of supplies and personnel. In making the observations here, I cannot recommend anything more forcefully than maintaining and even adding to our local emergency medical and public health expertise.

Whereas people in general want access to technology when needed and consider modern healthcare practices to be among the important advances of the era, most pundits agree that preventing health problems at the outset rather than expecting interventions to solve them is prudent. We can't always outfox our genetic heritage or stop an accident that causes broken bones or some other injury, but there is much we can do to prevent other kinds of injury and illness.

The coupling of preventive and primary care may be the best use of medical resources in the coming age. Promoting the synergies between the two models acknowledges the strength of each while encouraging their interdependence. Hierarchies in any social structure are to be expected, but the hierarchies of medicine have been bad for health care. We will surely need more cooperation and collaboration when we have fewer natural resources; preventive care and primary care are ready allies, even now. In Tompkins County several well-respected primary care physicians and family nurse practitioners seek out collegial relationships with complementary and alternative providers, thereby producing on a local level the integrative medical model increasingly, albeit quietly, under construction all over the world.

Some current technologies may be adapted to energy descent or saved outright due to their utility. One such technology could add to the models of care available in a remote place like Tompkins County. Telemedicine, the use of telecommunications devices to transmit medical information, complete examinations, and conduct surgeries, among other things, has been used successfully in a wide range of care settings. Some teaching hospitals use the technology to extend teaching and learning opportunities to distant sites; some use it to make surgical and other procedures more widely available. For more than 15 years, a few home care and hospice agencies across the U.S. have used telemedicine to make more efficient use of nursing and ancillary services and to allow patients, nurses, and other providers to see one another and to communicate complicated situations without taking on the burden of extra home visits. As the internet becomes more robust and ubiquitous, it is easy to imagine that the current monitor and phone line set-up typically required for telemedicine will be transformed by greater adaptability without much more of an investment in or expectation of increased technology. As energy descent ensues, maintaining the infrastructure required to power the internet will be a multifaceted asset.

Today residents rely on local specialists or specialists in Syracuse, Rochester, New York City, and out-of-state medical centers for some of the more arcane problems related to health status. Both energy descent and climate change will make travel to far-flung destinations difficult, costly, dangerous, and often impossible. Access via a screen may be the most we can expect when our local medical resources are not enough.

Finally, self care is the model health professionals of all stripes promote at the foundational level. Few "patients" can achieve self care, because once people become patients they're also sick and in some jeopardy of ever resuming a state of wellness. If as a community we aspire to knowing, protecting, and grooming our bodies and minds, we can be full partners in our tenancy here, which will make us all the more capable of managing other aspects of energy descent and climate change. For the purposes of realistic management in an energy-constrained world, self care includes knowing how to evaluate one's needs, adhering to a plan for achieving those needs, and being aware and capable of administering basic first aid, at a minimum.

CONCLUSION

Health care in the 21st century is a complex service requiring a complex set of skills. We can anticipate that aspects of the discipline will become more basic as energy descent and climate change progress. Residents can do much to prepare for altered expectations by learning concepts of basic care and by participating in planning for healthcare delivery in an energy-constrained environment. Supporting primary care and methods that lead to the prevention of illness, as well as the interdisciplinary model of integrative medicine, are helpful, proactive actions. Similarly, residents can provide oversight by insisting on the security of emergency and public health resources and by taking responsibility for the self care of their families.

REFERENCES

Alexander, C., S. Ishikawa, and M. Silverstein, *A Pattern Language* (New York: Oxford University Press, 1977), p. 255.

Bednarz, D., "Medicine After Oil," *Orion Magazine,* July/August 2007. Available at http://www.orionmagazine.org/index.php/articles/article/314/

Bednarz, D., "Energy and the Health Sciences: A Strategic Management Perspective," *Energy Bulletin,* August 8, 2008. Available at http://www.energybulletin.net/print/46146

Bissell, R., A. Bumbak, M. Levy, and P. Echebi, "Long-term Global Threat Assessment: Challenging New Roles for Emergency Managers," *Journal of Emergency Management,* Vol. 7, No. 1 (2009), pp. 19-37.

Chamberlain, S., *The Transition Timeline for a Local, Resilient Future* (Vermont: Chelsea Green Publishing, 2009).

Jeffrey, S., "How Peak Oil Will Affect Public Health," *International Journal of Cuban Studies* (June 16, 2008). Available at http://www.energybulletin.net/print/45750

Vision 2020: Final Report of the Addison County Conservation Congress. Available at http://www.acornvt.org/Documents/Vision2020.pdf

H

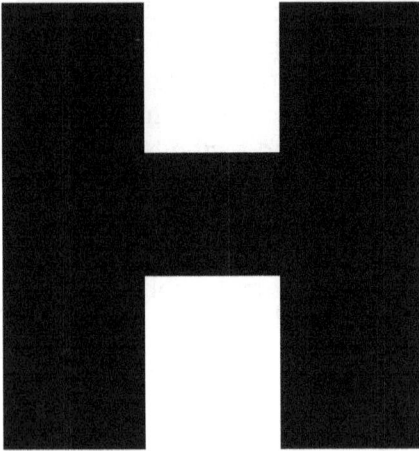

Health and Food Security

By Bethany Schroeder (January 2011)

INTRODUCTION

To many people, health is largely a matter of perspective. In the main, we subscribe to a working definition that includes feeling physically good; able to act and react according to some semblance of a reasonable self image; remaining fit in a passable manner; and weighing in at something near the insurance industry's norms.

Food security is another matter: some people describe food security as little more than being assured of the next meal, whereas others are unsatisfied with anything less than pantries full of canned and dried goods and well-stocked freezers. Members of disciplines as disparate as nutrition, planning and development, medicine, social justice law, and the armed services have considered the meaning and uses of the term with a view to overcoming the implied warning in its terminology.

Both health and food security are fraught with expectations at social, academic, and governmental/regulatory levels. Both are states of mind as well as physical conditions. Absent either, the human organism eventually dies. In short, health and food security are necessary to life—all life, and in the case of the present examination of the terms, most pointedly to human life. Health and food security are worth consideration because they are basic to life and because they have at all times in specific contexts existed in some imbalance. In general, when it comes to health and food security we expect much and plan all too little.

FOOD SECURITY VERSUS FOOD INSECURITY

Depending on the audience, experts have defined food security in formal and informal ways. In 1996, participants at the World Food Summit identified the presence of food security as in effect "when all people at all times have access to sufficient, safe, nutritious food to maintain a healthy and active life."[1] Participants also emphasized the combined requirements of being able to find and afford both nutritious food and food that meets an individual's preferences.[2] According to the Bureau of Public Affairs, it is thought across the globe that, quite simply, people are food secure when they can find and pay for food. Under this rubric, families are food secure when the members neither experience hunger nor fear starvation.[3] Furthermore, people with ethnic traditions and socio-religious mandates require that food be culturally appropriate. Many will refuse foods—even when hungry, even when in the midst of a food shortage —that fail to meet their expectations.[4] At least one local source, the Community Food Security Coalition, maintains that "community food security is a condition in which all community residents obtain a safe, culturally acceptable, nutritionally adequate diet through a sustainable food system that maximizes community self-reliance and social justice."[5]

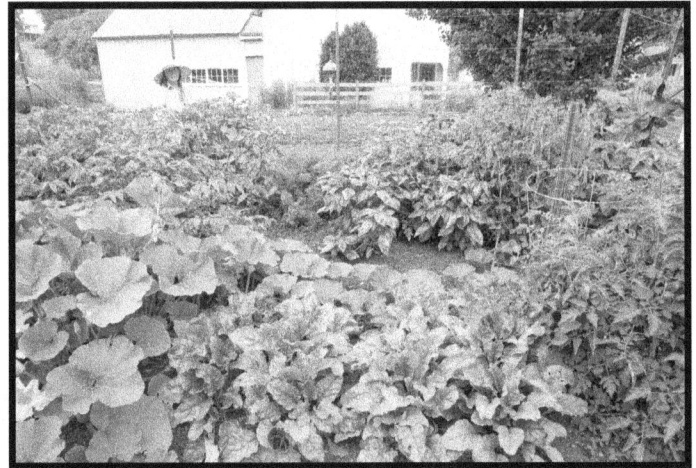

The author, in search of food security

Regarding the relationship between health status and food security, it may be sufficient to define good health as the ability to withstand the effects of exposure to illness and injury. The connection between nutritious food and health status is, from this perspective, fundamental, whether or not innate. Leaving aside the question of how to educate people to make healthy and nutritious choices, assuring access and affordability becomes a matter of public policy and the generous application of social support.

Also worth noting is the counter-intuitive notion of wide-spread hunger and food insecurity in the presence

---

1   http://www.who.int/trade/glossary/story028/en/

2   Nutritional research has shown that social, religious, and personal food preferences play a significant role in maintaining appetite, ultimately influencing the quality of an individual's diet.

3   http://www.state.gov/s/globalfoodsecurity/129952.htm#

4   http://www.who.int/trade/glossary/story028/en/

5   http://groundswell-ithaca.blogspot.com/2010/12/working-toward-food-security-in-ithaca.html

of abundance. Inequalities in distribution combined with general and pervasive poverty and a lack of knowledge about food preferences and prohibitions can result in food insecurity so endemic that neither individuals nor communities can overcome barriers to supply and access adequate to mitigate the problem.

In the past couple of decades, the terms and circumstances of food insecurity have been the subjects of increasing scrutiny. Citing 1990 research findings, the USDA describes food insecurity as ". . . limited or uncertain availability of nutritionally adequate and safe foods or limited or uncertain ability to acquire acceptable foods in socially acceptable ways." [6] What is more, the conditions associated with food insecurity are just those that we expect will result from declines in the availability of energy and the subsequent threats to the status of human health.

HUNGER, FOOD INSECURITY, AND THE EFFECTS ON HEALTH

Until recently, the absence or presence of hunger was the primary measurement by which many experts assessed food security as it applies to an individual's well-being. Without minimizing the significance of hunger, researchers recognized that hunger in a household might be an inconsistent problem and might apply primarily to one or more persons without being true of the entire household. Wanting to understand the role of hunger as it relates to food insecurity, researchers and policy makers began to think about food security or insecurity in the larger context of the community and the availability of food in general. Questions that routinely arose included the following:

What are the circumstances of hunger in a household?

Who, in a household, experiences hunger, and why?

What are the effects of hunger in a household?

What is required to relieve hunger, both temporarily and permanently?

Such inquiries found that hunger is typically the result of inadequate resources to obtain food but can exist when food choices are limited, too. Hunger often affects select adults who may ration food for more vulnerable members of the household. In the presence of food insecurity, hunger can affect everyone, especially the very young and the very old. Effects can include periodic hunger and the potential to develop food insecurity, if a lack of resources to acquire food or the unavailability of food is the cause of hunger. To achieve short- and long-term improvements in relieving routine or chronic hunger accompanied by food insecurity requires that planners, leaders, farmers and other food producers, just to name a few invested parties, develop a systemic understanding of the problem.

As a result of this and associated research, the USDA in particular altered its use of terms related to hunger and food insecurity, and has continued to look for refinements in ways of categorizing and addressing both phenomena. Germane to this TCLocal article is the realization by USDA and others that the understanding of hunger deriving from food insecurity ". . . results in discomfort, illness, weakness, or pain that goes beyond the usual uneasy sensation [of hunger]."[7] Especially during the past two decades of discussion and investigation, policy makers and those responsible for conducting research and the instruction of the next generation of field and university researchers and educators have come to appreciate the connection between food insecurity and the conditions, manifestations, and ramifications of ill health. Among other things, the implication is that hunger, in addition to being a symptom of food insecurity, is also a part of the panoply of conditions that signal compromised health status.

Undernourishment and malnutrition are two conditions widely agreed to be the results of hunger and food insecurity. Among children, conditions that can coincide with the latter include weight loss, fatigue, stunting of growth, and frequent colds. Studies have shown that undernourished pregnant women are more likely to bear babies with low birth weight, and the babies are then more likely to experience developmental delays that can lead to learning problems.[8]

Iron deficiency anemia is also common among hungry and food insecure children on one end of the spectrum and older adults on the other. In children, the condition can cause delays in development and learning. Children with iron deficiency anemia are also more susceptible to the effects of lead poisoning. In people of every age group, iron deficiency anemia can cause fatigue, weakness, shortness of breath, and irregular heart rhythms, among other symptoms.[9]

Moreover, hunger and food insecurity worsen the effects of all diseases and can accelerate degenerative conditions, especially among the elderly. Hunger and food insecurity create psychological responses such as anxiety, hostility, and negative perceptions of self-worth.[10] In an energy- and resource-constrained world, diseases like malaria, HIV/AIDS, dengue fever, and other infectious conditions from distant places, which experts anticipate will migrate in reaction to changes in weather patterns, can be expected to become more prevalent. More frequent incidents of these and other opportunistic diseases

6 http://www.ers.usda.gov/Briefing/Food Security/measurement.htm

7 http://www.ers.usda.gov/Briefing/Food Security/measurement.htm

8 http://www.frac.org/html/hunger_in_the_us/health.html

9 http://www.frac.org/html/hunger_in_the_us/health.html

10 http://www.frac.org/html/hunger_in_the_us/health.html

are likely to be reported, resulting in the potential to overburden the ability of any medical or public health system that tries to address the problem(s).[11]

LOCAL CONSIDERATIONS IN COMBATING HUNGER AND FOOD INSECURITY

In an energy-constrained future, such as TCLocal envisions in the next 10 to 20 years,[12] food insecurity and its consequences are expected to be increasingly common. The combined pressures of a larger population, climate change, reduction in the adjuvant energy required to grow food as well as the increased cost of such energy, and the potential for reduced or altered water resources could all create the environmental circumstances that lead to food insecurity. In fact, simply based on a growing population with the means to purchase choice foods, the demand for food could increase by as much as 50 percent by 2030. On the other hand, researchers speculate that increased demand and falling productivity could create widespread hunger and food insecurity, especially in the poorest communities of the world. All over the world, taking a preventive approach to food insecurity will require that we improve agricultural productivity and make access to markets easier.[13]

The outlook for our region is likely to be similar to that of the rest of the Northern Hemisphere, if not the world. The good news is that many of the residents of Tompkins County have developed an appreciation for the need to husband resources, as well as some of the skills to be effective at the practice. Locally, educators in well-established and informal venues alike have focused on the connection between promoting food security in combination with supporting good health, underscoring that each facilitates the other.

Assessing food security on a local level at this juncture with a view to predicting the potential for future changes will allow for planning and intervention. For example, according to 2009 statistics regarding the perception of hunger in Tompkins County, people across all income levels reported that the problem was widely evident. Twenty-three percent of respondents in the county's COMPASS survey said that having enough money to buy food was a problem in their own households. The use of local food pantries increased 30 percent between 2003 and 2008. Food stamp use also in-creased during the same period, with 4,223 households reporting participation in this subsidized food program in 2008 versus 2,288 households participating in 2003. Between 2001 and 2007, increases in reduced-fee or free lunches were noted among school children, with a quarter or more of all students in Groton, Dryden, Ithaca, and Newfield receiving support in the purchase of their meals.[14] Thus even in Tompkins County, where the standard of living is widely thought to be above average, a notable number of households experience hunger and food insecurity.

Though more can be accomplished, much is being done to address the problems associated with declines in health and food security. The services of agencies like the Department of Social Services, Catholic Charities, TC Action, the Red Cross, FoodNet, and others directly address local problems and enjoy an overall reputation for effectiveness. At the same time, the notable array of local Community Supported Agriculture seasonal options, the variety of U-pick and share farms in our area, the small and large market gardens, and the many agencies and local programs that educate people about how to use and preserve food have increased the general awareness of the need to address food security in Tompkins County.

A short list of local access-oriented programs includes emergency food services through Loaves & Fishes and the Salvation Army; the United Way's Food Pantry Garden in Brooktondale; the school district's Fresh Food and Vegetable program, which serves elementary age children; and assistance to childcare providers, parents, and pregnant teens through the Child Development Council, just to name a few. The Human Services Coalition's Information and Referral program and 2-1-1 Connect have also been helpful in directing people to much needed resources, including food resources.[15] Web-based support is available through the Community Cooperative Extension and the locally developed websites of Prepared Tompkins, IthaCan, and Harvestation. As is true in many communities across the U.S., in Tompkins County the internet has the capacity to connect people with resources by way of specific mail lists that promote local activities and community solutions to many problems, including hunger, food insecurity, and the consequences for health.

Increasing awareness of the existence of or potential for hunger among our neighbors and friends has spurred local efforts to find immediate relief. Although considered an unsanctioned method of food collection in some parts of the Western world, gleaning is not uncommon in communities across the U.S. In this region, grassroots efforts to serve and protect the poor among us

11 R. Bissell, A. Bumbak, M. Levy, and P. Echebi, "Long-term Global Threat Assessment: Challenging New Roles for Emergency Managers," *Journal of Emergency Management*, Vol. 7, No. 1 (2009), pp. 19-37.

12 http://tclocal.org/2010/10/outlook_for_liquid_fuels.html

13 http://www.usda.gov/wps/portal/usda/!ut/p/c4/04_SB8K8xLL-M9MSSzPy8xBz9CP0os_gAC9-wMJ8QY0MDpxBDA09nXw9D-FxcXQ-cAA_2CbEdFAEUOjoE!/?navid=FOOD_SECURITY&parentnav=FOOD_NUTRITION&navtype=RT

14 http://uwtc.org/compass-ii-20-social-issues-key-findings

15 http://uwtc.org/compass-ii-20-social-issues-key-findings

have been responsible for large local gleaning projects, frequently announced on the mailing lists of Sustainable Tompkins and the Finger Lakes Permaculture Institute, among others.

Local food security is also promoted by community gardens, where area residents not only grow food for their tables but practice prevention and health promotion in the act of working outside. Many local groups, including TCLocal, the Level Green Institute, and Sustainable Tompkins, have called for the development of this readily available solution to the problem.

LOCAL OPTIONS FOR ENHANCING HEALTH AND FOOD SECURITY

At an evening meeting of farmers and others interested in issues related to local food production, one of the farmers responded to the question, "Can the farmers in Tompkins County feed the population here?" with, "No. We can probably provide just 20 percent of the needs of the local community." Important in this anecdote is (1) the farmer's frank assessment and (2) that the question arose just four years ago.

Others have asked whether New York State can feed itself.[16] Indeed, in a time of energy descent, when fewer resources are available to grow and transport food, the potential for growing food closer to home, as well as recruiting and supporting local growers, may be among the most important questions to ask. In addition to assessing how much food production is possible locally, planners, growers, and area residents should consider the ways in which each might contribute to the solution rather than merely being part of the problem.

For starters, every yard and container has the capacity to be a food garden of one kind or another. Today the activity might primarily focus on cultivating the skills to grow food, whereas future circumstances may require skills honed to fill the table and the larder. Spending time outside in the garden encourages bone density through the absorption of vitamins. It also helps to build muscles and to keep the body fit and healthy. While not everyone likes to work in the garden, most of us like to eat. Learning to think about food production as a civic responsibility has historical contexts all over the world, as much in Tompkins County as anywhere else.

Legislators at all levels of government could help more of us to be producers rather than only consumers of food. Suspending or discontinuing ordinances that restrict farming, gardening, and tree-crop production could encourage more participation in the food economy, most likely at the informal level. Whether considering food for sale, barter, or personal consumption, reducing the unnecessary barriers to food production that inhibit growers is the first step in ensuring that everyone has enough to eat. Rethinking area ordinances about the management of food and food systems will be necessary to enhancing health and food security in an energy-constrained world. Considerations of what constitutes agricultural land, who can hold it, and how it's taxed should be topics of discussion at county, town, and city levels of local government.

In general, Tompkins County has an abundance of fertile, versatile land and adequate water supplies to promote the growth of every manner of food that can be produced in this climate. Increasingly significant in the study of agricultural techniques are nutritional outcomes, depending on the quality of soil and its augmentation. Despite many studies and much debate, the jury remains undecided about the relative value of organic versus conventional methods of soil management for the sake of healthy nutritional impacts.[17] Nonetheless, researchers do agree that organic methods produce less environmental stress. At the very least, the absence of additives, typically derived from natural gas, commends organic techniques to the small farmer or gardener in circumstances of energy descent. No matter which methods we use to grow food, we must thoughtfully manage the short- and long-term integrity of the soil if we want to help retain its best characteristics year after year.

So too must we be careful stewards of the region's ponds, creeks, and lakes. At the local level, the protection of all water resources is a matter directly related to health and food security. The public health department oversees the potability of water, relying on standards set at state and federal levels. Common sense and a basic understanding of interdependencies are enough to show that poor management of our water will affect whether we can grow adequate food, not to mention whether water supplies are safe for our consumption and for consumption by livestock.

Animal husbandry includes the allocation of important food resources, but the practice is presently defined and permitted according to economic standards that we, under circumstances of reduced access to energy, cannot hope to sustain. Owning a cow or a flock of chickens, for example, may not be necessary to every family, yet the availability of milk and eggs locally sold (or shared) and produced might well come to be viewed as a necessary feature of community life.

At the same time, assuring that those who work to grow food, whether formally or informally, have access to hygienic resources makes good sense from the per-

---

16 See the article beginning on page 10.

17 http://www.foodsecurity.org/FPC/council.html. One of the best Websites I found while completing the background reading for this article, the site is rich in subjects that range from agronomy to wildcrafting, from vitamin deficiencies to nutritional variances among indigenous peoples.

spectives of safeguarding the talent and skill necessary to effective farming and gardening and to the quality of our food at its source. People need bathrooms and sinks or other hand washing options, especially options that don't contribute more trash to already overburdened landfills or the use of supplies made from oil or natural gas. We could make facilities more widely available near gardens and farms, and we could manage them locally.

Discussions at NOFA conferences and other similar meetings are reportedly well attended, exhibiting the kind of regional knowledge and sensitivity to local issues that supports asking important questions about food issues and promotes success in approaches to planning that address those issues. In particular, food policy councils, frequently made up of interested professionals, community members, farmers, vendors, and legislators, have proven to be useful in some communities in helping to organize the selection, production, and distribution of food.[18] As noted earlier in this article, a loose coalition of food experts and community organizers in Tompkins County has lately convened to discuss the possibility of an area food council. Among others, issues explored included, first, the activities helpful to improving the local food system via a food policy council, and second, the necessary resources and commitment needed for success.

In this article, I have described just a few considerations related to health and food security. I hope that others will follow up my work with a deeper and more expert examination of the issues. In addition to adhering to the principles that guide TCLocal in its goal of understanding how residents might operate with fewer resources and more sustainable approaches to development, I recommend that we examine first principles of fair access, fair use, and fair expectations regarding health and food security. A healthy, integrated, and self-aware community must learn how to share resources, recognizing that the whole is only as strong as its weakest part.

---

18 http://www.foodsecurity.org/FPC/council.html

B

Burning Transitions

HOW PLANNED, LOCALIZED, SUSTAINABLE NON-FOOD BIOMASS UTILIZATION CAN HELP EASE ENERGY DESCENT AND MITIGATE GLOBAL CLIMATE CHANGE

by Krys Cail (October 2009)

INTRODUCTION

This article provides a framework for considering the socio-economic structural changes that can lead to a different, more stable, and more sustainable local market for heating fuel and electrical energy.

The use of combustion for heat and power is an established and developed technology, while the successful social balancing of environmental and ecological costs with short-run economic benefit is a new, and daunting, challenge. The change, or transition, needed to use the locally available resource of non-food woody and grassy biomass to help solve current energy problems is socio-economic change, not technical innovation. We can supplant at least some current fossil fuel use with the more carbon-neutral combustion of earth surface harvested feedstocks using current technology. Nonfood biomass direct combustion[1] can be undertaken in a localized context. We can take an enlightened approach to the sustainable management of feedstock planting, growing, and harvesting, energy-efficient processing, complete and clean burning, and ash recycling. Developing such a system also offers a means of developing the alternative commercial channels necessary to move the Tompkins County area to a future of heat and energy production that is not just more environmentally friendly, but also more economically insulated, or decoupled, from the gyrations of the world oil market in a time of post-peak oil.

Other current and emerging heat and power technologies, such as solar, wind, geothermal, and small-scale hydro are "greener" forms of alternative energy and may be our future mainstays. However, in biomass-rich locations like Tompkins County, the economic attraction of biomass as an affordable substitute for fossil fuels will ensure that it will come into commercial use as oil and other energy commodities rise in price. If the development of biomass energy is controlled by the current energy industry, large energy companies will guard their market share by organizing only large-scale markets, even in situations where energy efficiency favors smaller, more localized scale. Conversely, building localized commercial structures to sell nonfood biomass-generated heat and electrical energy could feasibly provide a template for the effective investment in and commercialization of localized energy from other, greener sources in the future.

The kind of community development that allows areas the size of Tompkins County to become more energy self-reliant—"import substitution" for the energy products of the fossil fuel industry—can accomplish the twin goals of creating green jobs and modeling the kind of less global, more local commercial/economic interactions that are referred to as relocalization. Relocalization of energy provision is a necessary response to energy descent; accomplishing this using tested community development practices will ensure better success in the required transition.

THE FIRST TWO BURNING TRANSITIONS

Combustion (fire), used as a tool, was a major human cultural advance, and perhaps helped our species to evolve. In his recent book, *Catching Fire: How Cooking Made Us Human*,[2] Richard Wrangman, a Harvard University biological anthropologist, postulates that the taming of fire, and its use to cook food, was the key tool-using event that allowed human evolution to proceed from pre-human hominid to modern humankind. He postulates that cooked food allowed us to divert calories from chewing to growing larger brains.

The centrality of fire to the establishment of human society is also evidenced in religions and belief systems worldwide. One classic rendition is the myth of Prometheus, the champion of humankind who was said to have stolen fire for use by mortals from the immortal gods.

From ancient times up until the Industrial Revolution, humans used combustion sustainably, with only localized or regional instances of deforestation.[3] Early burning was carbon-neutral as far as the earth's atmosphere was concerned.

---

1 "Direct combustion" refers to biomass burned as a solid fuel, not a liquid or gas fuel product or fuel additive.

2 New York: Basic Books, 2009.

3 Localized or regional deforestation should not be underestimated in its capacity to decimate human, animal, and plant communities, including driving some species to extinction. It does not, however, represent a pattern of world-wide changes, despite its severe impact on circumscribed areas.

Some primitive peoples did set massive fires. For instance, Plains Indians used prairie fires to stampede buffalo over cliffs; Tompkins County's first peoples probably (like New England's natives) routinely burned the forest understory to make for easier hunting access;[4] and innumerable horseback European raiders ransacked and ruined settled villages with fire—as Revolutionary War General Sullivan did here in the Finger Lakes. These combustion materials were already a part of the earth surface/atmosphere carbon exchange. In geologic/atmospheric time, even very big surface fires are just blips. The carbon released into the atmosphere would have otherwise been added shortly anyway through decomposition. It was the Industrial Revolution and the use of first coal, and then oil and natural gas, that began the process of unbalancing the planet's atmospheric carbon load by making use of the carbon stores of former eons, previously safely buried underground. This led to both global climate change, and to the depletion of easily extractable in-ground carbon sources we speak of as peak oil and energy descent.[5]

The first "burning transition," then, was the Prometheus transition. This transition changed humankind (if you don't believe Richard Wrangman that it changed our evolution, you must at least concede that it drastically altered our culture). The Prometheus transition enabled the development of agriculture and led to deforestation in a few subcontinental areas. But the second burning transition—and the advent of the steam and internal combustion engines of the Industrial Revolution—resulted eventually in major land and sea transformation and widespread ecosystem and climatic change. The first burning transition changed humankind, while the second burning transition changed the planet. Each burning transition also markedly changed the socio-economic systems that people used to regularize and control the commercial and familial relationships that provide us essentials such as heat in cold weather, food, and, after the second burning transition, electric power.

PLANNING A THIRD BURNING TRANSITION

Technological optimism about alternative fuel development usually focuses on replacing combustion of "dirty" fuels with combustion of "clean" fuels, while leaving the production and distribution systems for liquid and gaseous fuels and electrical power in its heritage configuration. That configuration is controlled by some of the most powerful international corporations on earth—oil and gas developing, refining, and shipping companies, electrical utilities, and coal mining and shipping companies. These actors have a vested interest in seeing that the socio-economic systems of the future do not deviate too much from those of the past, ensuring these corporations continued market share. Is that to our advantage?

Is the needed change limited to a substitution of one fuel for another, one feedstock for another, or one power source for another, with no substantive change to social, industrial, political, or economic institutions? Or is a more substantive transition needed? Will social and economic change follow technology, or will we invent and popularize only the technologies our social and economic systems predispose us to aim toward?

"Local planning for sustainable use of local resources" is the basis of egalitarian post-colonial social and economic development. It is also the key to the development of a third, socioeconomic/cultural burning transition. Rather than assume an international market in energy as a given and hope for technological fixes, we should focus in the third burning transition on the relocalization of systems of sourcing, producing, and distributing heat and power. In that context, the on-going technological development can be decoupled from the economic fortunes of transnational corporations that are difficult to call to account on environmental effects in any particular place. A different kind of optimism about confronting the challenges of global climate change and peak oil can be envisioned, one in which the needed change in socio-economic structures is the direct goal, in order to accomplish the most efficient and environmentally-sound use of energy *within current technological and environmental limits.* This might then be followed by additional technological advancement, as needed and affordable—perhaps even a Solar or Geothermal transition that makes burning itself unnecessary. However, those possibilities are too far away for a complete transition right now, and right now is when global climate change must be addressed. Rather than trust humanity's on-going scientific and technological innovation to "come up with something" that will make unfettered world markets in energy able to function within environmental limits, this optimism postulates that human communities can learn to balance their own energy needs with the sustainability of their own environments through socio-economic or socio-political progress.

The third burning transition is, in essence, a relocalization of energy production and an implementation of the household and commercial structures needed to manage more local production and consumption of energy, one that brings the source and use points of energy geographically closer together. This is a transition that requires no new or special technological development,

---

4   Cronon, William. *Changes in the Land, Revised Edition: Indians, Colonists, and the Ecology of New England.* New York: Hill and Wang, 1983.

5   Biomass/ethanol/biodiesel schemes dependent upon intensively cultivated food crops like soy or corn fail to break the connection between the oil market and alternative fuel if a system of petrochemical input dependent agriculture is used. They also raise grave ethical concerns, commonly referred to as the "food-fuel controversy."

but rather advancements in business form development and industrial design, including business and consumer combustion equipment and new approaches to the design of district heating and electrical power grids.

THE NEED FOR A LOCAL APPROACH

Localities differ in what kind of resources they have available to produce heat and power. Thus far, most research and development in the area of biomass use as an alternative energy feedstock has used a non-localized model. Raw biomass is generally first converted into liquid fuel (both corn-based and cellulosic ethanol are liquid fuels), and then distributed via pipeline, tanker, and tanker truck, similar to petroleum. Or, alternatively, biomass is burned directly, but the resulting heat is made into electric energy and distributed far and wide on the electric grid. *Both of these models contain large distributional inefficiencies.*[6]

Government subsidies for one form of fuel over another can have unforeseen effects. Often, governments subsidize use of "cleaner" or more carbon-neutral fuels or combustion equipment via a direct consumer subsidy, such as a tax credit, or an indirect subsidy, such as a producer tax break or capital investment in production plant and equipment. Corn ethanol—an alternative fuel that even its promoters are now seeing as a "transitional" alternative fuel—is an example of how governmental enthusiasm for jobs, plant, and equipment in every legislator's district can result in a glut of relatively expensive alternative fuel production in remote areas, with little hope of export at a profit in the face of price variation in the oil markets, where the product competes directly.

Some European governments have backed the development of small-scale solid-fuel biomass combustion, from pellet stoves to wood-chip furnaces to multi-fuel-burning combustion units and ultra-efficient gasification boilers that power electric generators as well as district heating grids. While this has led to much more widespread adoption of the technologies than in the US, there are still some perverse global-market effects. The governmental support for wood pellet burning in Northern Europe (direct consumer subsidies for pellet stoves, for instance) has resulted in the US market for wood pellets being significantly impacted by European demand: shortages of wood pellets in both the US and Europe in recent years have been blamed, in part, on the fact that most wood pellets produced in the US are shipped, under contract, to Europe, rather than available

for growing domestic use.[7] If the domestic demand for wood pellets rises because fuel oil rises significantly in price, manufacturers can't satisfy it, and resulting shortages drive up wood pellet prices in tandem with fuel oil prices.

Government support for the development of green energy is surely needed. But, as illustrated above, direct support for particular technologies can have perverse outcomes, when, in the real world, the variable and uncontrollable price of oil interferes with orderly marketing of the product as a substitute for the fuel and power sources people are accustomed to using. For that reason, localized community-controlled energy development for heat and electricity is preferable, as it can reasonably allow a community or geographic region to claim its own energy resources and begin to decouple its energy costs from the world oil market. In addition, as is the case with consumers developing commercial relationships with their local farmers, a measure of consumer loyalty and flexibility can be gained by localizing the transaction.

LOCAL RESOURCES: PRIME DETERMINANT OF APPROPRIATE COMBUSTION FEEDSTOCKS

The third burning transition will look different in different locations. Relocalization offers the opportunity for each region or locality to assess what underutilized or sustainably developable resources it possesses, as well as what market power its heat and energy consumers represent. The skills and resources of local people must be accounted for, as well as underutilized natural resources and plant and equipment in the built environment. This assessment of resources can be done as a part of a tried and true methodology of community and economic development—Asset-based Community Development.[8] An asset-based approach to community development allows for customizing programmatic goals to highlight natural resources, human capital agglomerations, and other local conditions that will make one form of biomass more feasible to use as a feedstock for combustion than another.

The local foods movement has made some use of the phrase "Eat your landscape." The idea is that, by engaging in an ongoing direct involvement in growing food (gardening or CSA working membership) or direct-

---

6 While current average distributional losses for electrical energy are in the range of seven percent, biomass resources, like solar resources, may be located at a greater distance from urbanized areas than existing power plants, resulting in even larger distributional losses or larger amounts of transportation energy to move the raw material closer to the point of use.

7 More on the international market volatility of wood pellets is available in the *Renewable Energy World* magazine article "Time for Stability: An Update on International Wood Pellet Markets," Feb. 4, 2008. Available at http://www.renewableenergyworld.com/rea/news/article/2008/02/time-for-stability-an-update-on-international-wood-pellet-markets-51584

8 See The Asset-based Community Development Institute at http://www.abcdinstitute.org/ or Wikipedia on Asset-based Community Development at http://en.wikipedia.org/wiki/Asset-Based_Community_Development

from-the-local-farmer commercial interaction with a manager of food producing lands in your locale ("landscape"), one can exercise, in common with one's neighbors, some influence over what kind of a landscape it is now and in future. The goal is and use that is environmentally sound, sustainable, and provides a living wage to those who manage and work the land.

A similar approach can be taken to the orderly and sustainable harvest and cultivation of biomass for combustion in place of oil, gas, and, especially, coal. Although these fuel substitutions are not the ultimate long-term solutions to our energy problems, they do offer us a mechanism for developing the distributed, local commercial interactions that can and will set the stage for the development of more long-term sustainable energy systems. In much of Tompkins County, for instance, woody and grassy biomass may be available for use as a combustion fuel, but the commercial infrastructure to sustainably and profitably grow, harvest, and process that biomass needs to be developed. Without a community development effort in this area, woodlands and pastures in Tompkins County will continue to fall into an unused and unmanaged condition that does not allow for optimum carbon sequestration and invasive plant control and does not support the development of local energy and green jobs.

In Tompkins County, most of the underutilized resource is privately owned forested or pasture/hay land that is minimally managed and, in some cases, is becoming overgrown with invasive brush species. The following chart shows the acreage of various types of landcover in the county.

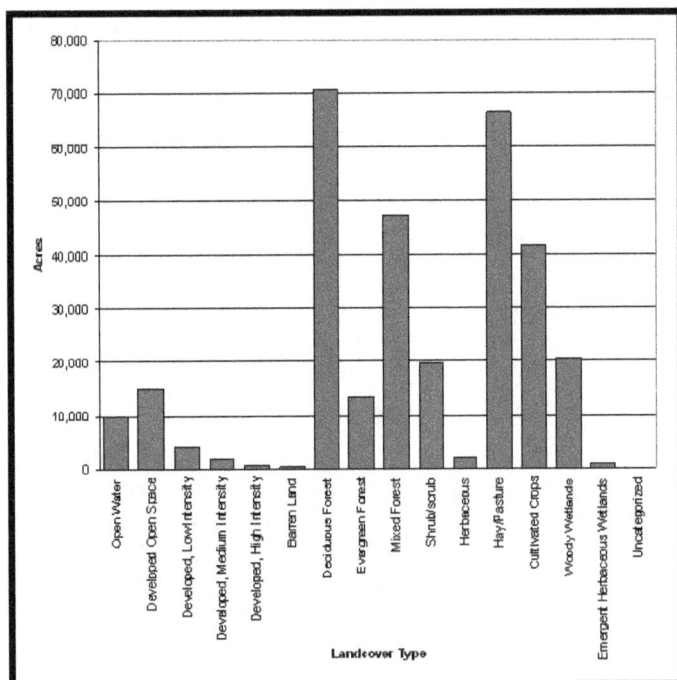

Figure 1. Tompkins County land types

In the map of Tompkins County on the next page,[9] land devoted to agriculture is rendered as white, land in forest or natural cover is rendered as medium grey, and all other land uses are rendered as black. This shows a pattern of land use that conforms to topography: the northern portion of the county, which is composed of flatter land and relatively more of the better soils for agricultural use, has a greater percentage of acreage in cultivated cropland and pasture, while the southern, hillier portion of the County is more densely wooded.

ORGANIZING FOR LOCAL ENERGY PRODUCTION AND CONSUMPTION OF BIOMASS

"Eating your landscape" implies sustainability. A bountiful landscape might continue to provide food over decades, centuries, even millennia if it were properly managed and husbanded. "Burn your landscape" has none of the overtones of sustainability—it seems, rather, cataclysmic: a landscape devoid of living things.

There are other options, however. An actively managed forest or hayfield can continue to produce biomass for combustion purposes over a long period of time if attention to the ecosystem allows for the return of depleted soil nutrients through ash spreading and the building of fertility through support of various plant and animal communities. Woodlands actively managed for sustainable harvest of woody biomass could provide plant and animal habitat, sequester carbon, and produce some hardwood lumber as well. The key here is *the way in which natural resource lands are managed.* Under some systems of management, carbon sequestration and selection to impede the advance of invasive species are optimized, creating a forest that is more hospitable to native flora and fauna and more able to ameliorate the excess atmospheric carbon than the previous unmanaged woodland. However, such management systems are not the most economically viable under current market structures.

Current economic structures, if left unchecked, could cause cataclysmic environmental damage as harvested biomass becomes less costly than oil. Clear-cutting woodlands, while devastating to natural communities and water quality, is the cheap way to amass a large tonnage of biomass in an area like Tompkins County. Utility companies buy wood-chip tonnage to co-fire with coal from low bidders, developing an industry built around mechanized, invasive forest destruction. Environmental regulation has proven to be a weak tool for controlling industries that have a market incentive to use forests or grasslands as a short-term, rather than permanent, resource. An example is the Catalyst

9 Adapted from
http://www.tompkins-co.org/gis/maps/pdfs/LULC2000.pdf

Energy/Treesource Solutions biomass aggregation facility in nearby Burdett, Schuyler County, which is offering loggers one low price for biomass tonnage to be used as wood chips to heat and power the US Salt plant in Watkins Glen.

On an individual scale, landowners who use firewood for heat are likely to take the long view of their investment in their land and do their best to manage their woods to maintain sustained production as well as multi-functionality (use of the woodlands for additional purposes, such as wildlife habitat, hunting, nature appreciation, privacy). When surveyed, owners of rural acreage in Tompkins County were amenable to seeing their underutilized parcels of land produce an income stream—but very few had either time or capital to devote to this.[10]

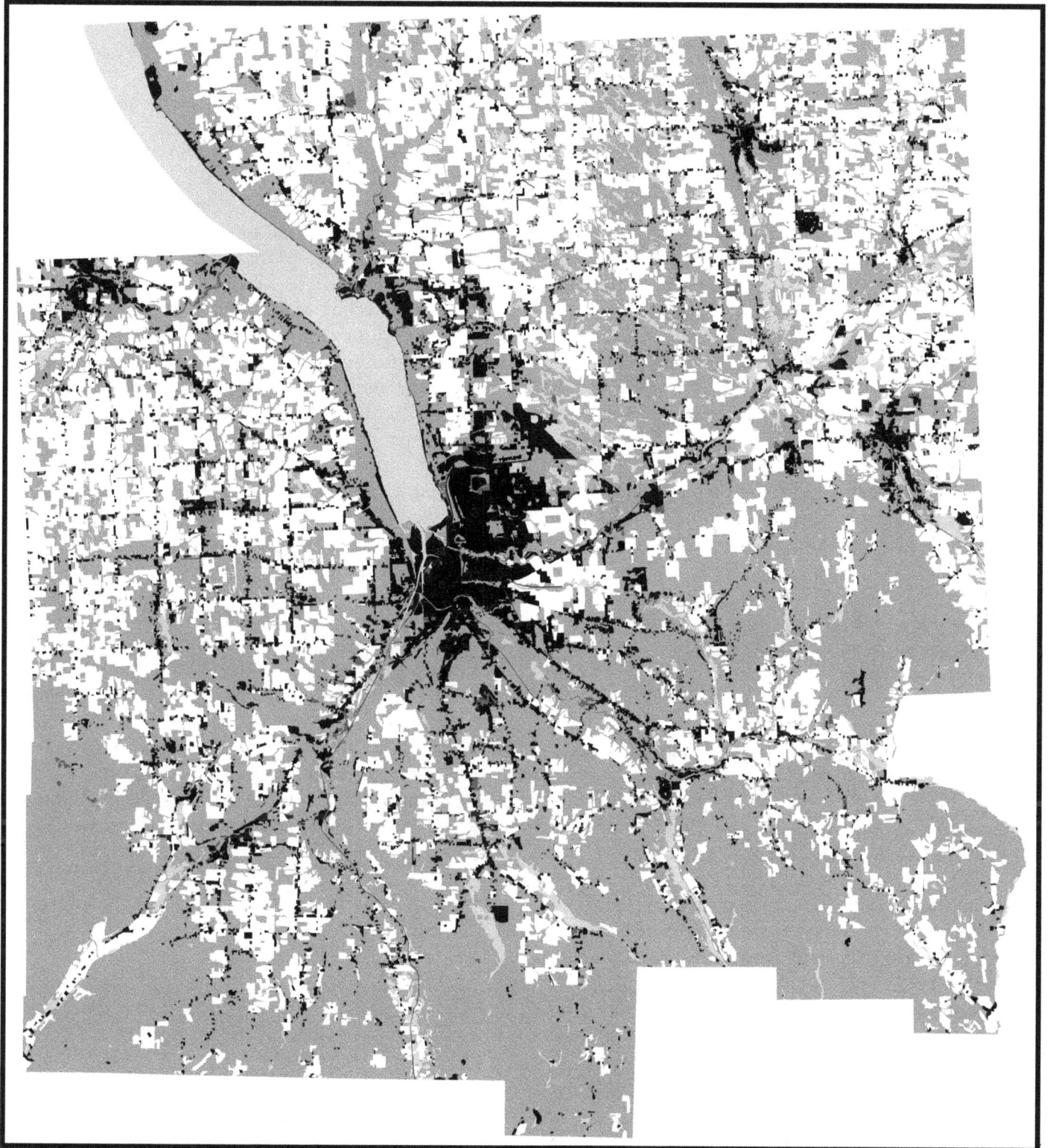

Figure 2. Tompkins County land use (white = agriculture, grey = forest or natural cover, black = all other land uses)

Several local initiatives in Tompkins County have sprung up to test structures that might become a part of a third burning transition here. In the Town of Danby, landowners have come together to market the biomass from their properties (as well as potentially other land-based products) as a group. This organization of owners of fallow fields and under-managed woodlots is based on the producer-coop configuration that has been successful in some agricultural areas.[11]

Another effort, spearheaded by Anthony Nekut, is intended to draw together investors and entrepreneurs with the purpose of developing a medium-scale pellet production facility in the county. Tony would like such a plant to have the capacity to palletize both woody and grassy biomass, and he envisions both local sourcing of biomass and local sales of pellets for home and business heating. *[Tony later contributed to the next article in this series, "Heating with Biomass in Tompkins County."—Ed.]*

A third approach to using biomass to supplant some of the fossil fuels used for home heating in Tompkins County is Abbot Development's initiative to develop Cornell University workforce housing on a Danish-style district heating model, with a combined heat and power plant as an integral feature of the development. This plan is currently in concept development stage, but it could easily be implemented if chosen by Cornell as the model for their new housing development. Again, the technology is available and ready to use; it is the commercial market structures that require some developmental attention to establish such a project in this country.

A fourth local project focuses on commercial combined heat and power along with a managed woody-biomass plantation scheme: RPM Ecosystems, a Dryden company involved in the production of fast-growing nursery stock for reforestation projects worldwide, has worked with Congressman Michael Arcuri to obtain federal funding for a demonstration project. The project involves a wood-fired combined heat and power plant that would provide heat for the greenhouses and offices of the nursery along with sufficient electrical power to operate the facility. Additionally, plantations of RPM Ecosystems trees would be established with a goal of producing some biomass along with some hardwood lumber while maximizing forest canopy (and carbon sequestration) throughout the growth and development of the tree farm.

One approach that is not currently in evidence in Tompkins County, but might be worth investigating, is the "CSE." CSE stands for "Community Supported Energy," and it is modeled on the successful CSA (Community Supported Agriculture) structure. This is something of a consumer cooperative: energy consumers that would like to use local resources to produce energy band together, and, through pooling investment funds, establish critical mass to bring a production facility on-line, which they pledge to support through their energy purchases. This model was first promoted by environmental advocate Greg Pahl, and has been tried with some success in Vermont.[12]

CONCLUSION

The above examples merely scratch the surface of possible structures for relocalizing our heat and energy markets. And the traditional approach should not be ignored, either: use of cordwood for home and business heating has increased markedly as fossil fuel prices increase and can be expected to continue to increase, particularly in rural areas of the county. More people now make a main business or a profitable sideline of harvesting firewood, or buy less fossil fuel because they harvest some firewood for their own use. Several local retail outlets and service businesses sell and/or install combustion equipment, and technology refinements have made cordwood burning cleaner and more efficient than it was in the past.

A third burning transition—based on community development and economic innovation—is needed if we are to avoid the worst potential effects of global climate change and post-peak-oil economic instability. In the first burning transition, fire changed humankind; in the second, humankind using fire changed the world until disaster threatened. In the third burning transition, humankind must organize new structures of production and exchange to socially contain the power that unlimited individual fire-use unleashes on the world, to protect both the species and the environment on which it depends. In the future, the structures so organized can be again transformed, in a fourth burning transition, to non-carbon-based feedstocks such as the sun's direct energy, geothermal heat, and wind and wave energy.

---

11 Begun as a project for Elizabeth Keokosky's masters degree in City and Regional Planning at Cornell University, this initiative has progressed to the point of establishing a local steering committee and is in the process of drawing up incorporation documents.

12 Pahl, Greg. *The Citizen-powered Energy Handbook: Community Solutions to a Global Crisis.* White River Junction, Vt.: Chelsea Green Publishing, March 2007. See also *Renewable Energy World* magazine, "Community-supported Energy Offers a Third Way," Greg Pahl, March 12, 2007. Available at http://www.renewableenergyworld.com/rea/news/article/2007/03/community-supported-energy-offers-a-third-way-47700

H

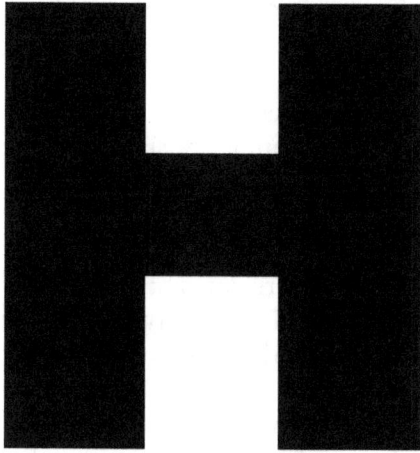

Heating with Biomass in Tompkins County

by Krys Cail and Tony Nekut (January 2010)

This article continues the discussion of heating with local biomass begun in "Burning Transitions." There it was noted that the best application for local biomass energy is combustion for space heating, possibly coupled with distributed CHP (combined heat and power) electricity generation, and that these technologies are, for the most part, already developed and available in the form of high-efficiency gasifying boilers and pellet stoves.

Work is required along the entire supply chain (growing, harvesting, processing, distribution, and utilization) if local biomass energy is going play a significant role in Tompkins County's energy future. The traditional economic stakeholders are a diverse group of mutually dependent players (landowners, loggers, foresters, farmers, manufacturers, fuel retailers, and consumers), each requiring commitment from the others to make the system work. Leadership and planning are essential to moving beyond gridlock by demonstrating how, through cooperation, everyone along the chain stands to benefit. Fortunately, there are a variety of case histories and other resources that have been developed in recent decades that render this demonstration somewhat easier.

Barring unforeseen breakthroughs in energy technology, it seems clear that this resource will indeed be developed. Local biomass is already cost competitive with fossil fuels for space heating, and its economic viability will only improve as fossil fuel prices continue to rise. The time has therefore arrived to begin development, because time will be required to build the needed infrastructure.

THE SCALE OF THE LOCAL BIOMASS DEVELOPMENT CHALLENGE

Every form of biomass yields about 16 million BTUs per dry ton when burned. Sustainable annual biomass productivity ranges from about 0.5 dry tons per acre for our local forests to 5 dry tons per acre for some locally suited energy crops. These productivities represent conversion efficiencies from solar radiant energy to stored chemical energy of about 0.1 to 1 percent. If half of Tompkins County's 300,000 acre land area were committed to growing biomass, the annual per capita energy production would range from about 12 to 120 million BTUs. (See the discussion of County land cover in the "Burning Transitions" article.) By comparison, current (2007) statewide annual per capita primary energy consumption is 219 million BTUs. In other words, the amount of biomass energy we could get from our land even in relatively rural Tompkins County would yield nowhere near our total energy needs.

Meeting our heating needs is another matter. Each household in the County uses about 100 million BTUs annually for water and space heating; this is about 43 million BTUs annually per capita—approaching the range of sustainable large-scale local production. Adding the wholesale implementation of residential energy efficiency measures would bring total heating energy self-sufficiency within reach. Ed Marx, Tompkins County Commissioner of Planning and Public Works, has been quoted as estimating that biomass could heat up to 40 percent of the homes in the county, or even more if homes were super-insulated.

THE BIOMASS HEATING GAP

Local biomass energy for heating has enormous potential benefits. It creates jobs, keeps money local, provides energy security, reduces CO_2 emissions (locally burned biomass is virtually carbon-neutral), increases carbon sequestration, slows fossil fuel depletion, improves forest and soil health, maintains rural land values, reduces development pressures, creates community ties, and raises community environmental awareness. But fewer than 5 percent of County homes are listed in census data as heated primarily with biomass (cordwood and pellets). For 2008, the Census Bureau's American Community Survey estimates the percentages shown in Figure 1 for heating the 37,749 occupied housing units of Tompkins County.

The apparent lack of interest in heating with wood shown by the 4.5 percent figure is partly an artifact of the way the data is gathered and partly due to active discouragement of wood heat by mortgage lenders and insurance companies.

Wood heat appliances do not enjoy wide acceptance by those who underwrite mortgages and insure homes.

Due to the perceived risk of fire, many underwriters of homeowners insurance will not insure properties with wood stoves. (Pellet stoves, which are less likely to cause chimney fires, are a bit more acceptable.) In particular, homes that include rental units—even if the home is also owner-occupied—are very difficult to insure if there is wood-burning equipment in use for heating and the insurer is aware of that fact.

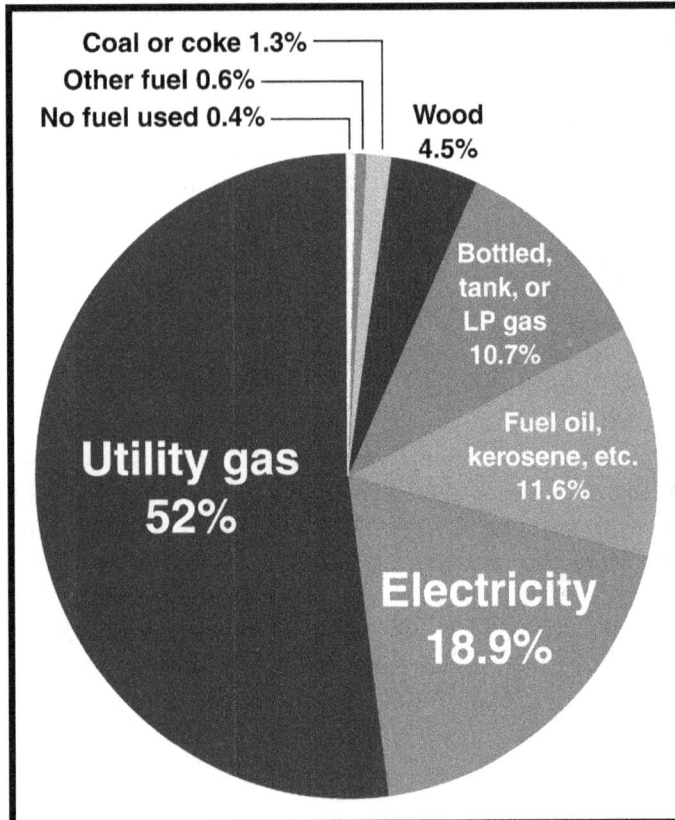

Figure 1. Sources of home heating in Tompkins County (2008 est.)

Of course, homeowners insurance is a requirement for any house that has a mortgage. But it is not just the reticence of homeowners insurance underwriters to insure homes with woodstoves that limits use of this technology; there is also a problem associated with wood heat when the lender packages the mortgage for resale on the secondary mortgage market. Despite the worldwide use of this simple technology, burning wood to heat a dwelling is perceived as too risky. Where woodstoves are a secondary, rather than primary, source of heat, this is overlooked. But a home that relies predominantly on wood for its heating source is a home whose purchase will be difficult to finance.

GAINING A MORE ACCURATE ESTIMATE OF LOCAL TRENDS

This situation—which makes it plausible to add wood heat as a secondary heat source, but difficult to rely on it as a primary heat source—helps explain why the census

figures for wood heat seem so low. This is a problem even in the decennial direct counts.

The between-decades estimates suffer from an additional problem: there's no mechanism for reporting a local trend. Households demographically similar to the various household types in Tompkins County are surveyed at various locations around the country, and the composite picture of their changes is applied to local households with similar characteristics. For instance, if upper-income professional couples with two or fewer children in the home were, for the most part, heating with natural gas across the US, there would be no mechanism for the American Community Survey to read a recent upswing in purchases of woodstoves and pellet stoves among local college and university faculty.

To get a better idea of what's really happening locally, we had to ask around. The results, while anecdotal, point to just such a trend.

The sales managers of both local woodstove/pellet stove retail outlets indicated that business has been steadily increasing throughout the decade, with particularly noticeable upswings in wood heat appliance purchases when other forms of fuel—particularly fuel oil—experienced price run-ups or price volatility. Both described their typical customers as college professors or other professionals interested in saving money and in helping to conserve non-renewable resources. While families living in more rural and suburban locations were the norm for local wood heat users at the beginning of the decade, the increasing popularity of pellet stoves has resulted in more urban families buying wood heating appliances.

A construction manager at Ithaca Neighborhood Housing Services, which administers a group of NYSERDA programs aimed at green energy for heating, concurred that more urban residents are choosing pellet stoves, and that the help available from NYSERDA resulted in more low- and moderate-income families being able to access affordable financing to add wood heat to their homes.

Another indication of the rising local popularity of wood heat of various sorts is the brisk business that fuel purveyors are doing in cordwood and pellets. The owner-operator of Finger Lakes Firewood, the largest local cordwood dealer, has purchased additional automated equipment to better clean and move his cordwood as his customer base has continued to expand. Ithaca Agway has been using its display sign to advertise pellets, and the Home Depot devoted as much space at the front door to sale-price wood pellets as to the snow blowers.

INDUSTRIAL USES OF WOOD HEAT

Wood heat is beginning to appear in local industrial operations, too. For example, US Salt in Watkins Glen is in the process of converting the heating of its large facility on Seneca Lake to biomass.

According to Len Boughton, an engineer with the firm who has been responsible for overseeing the construction and retrofitting, the system, after two years of work, is now in place and operational, but the switch to wood-based fuel will wait till March to allow troubleshooting during a season of less extreme heating demand.

Plant Manager Frank Pastore said that US Salt has contracted with TreeSource Solutions (http://treesourcesolutions.com/) to avoid the management burden of dealing with multiple suppliers. TreeSource is a wholly owned subsidiary of Catalyst Renewables (http://www.catalystrc.com/). Pastore said that he expected the bulk of the fuel to come from local sources, through the Wood Yard that TreeSource has established nearby in Burdett, but that he trusted the contractor to source wood fuel as appropriate in order to maintain a stable and affordable price.

BUYING AND SELLING BIOMASS IN BURDETT

The Wood Yard at the old railroad depot in the Village of Burdett was, in its last incarnation, a steel recycling facility, and many of the buildings are simply being re-used "as is"; the old depot itself is used as a scalehouse for weighing trucks. The facility includes a large, rambling lot with a gated entrance from State Route 79. The Yard was not officially open the day of our visit, but it's clear that the facility is used in a number of synergistic ways in addition to providing a means to weigh and store wood intended for use as biomass fuel.

A recently constructed pole barn houses a portable bandsaw mill, and some rough-milled lumber showed that the facility is in active use. A large pile of logs awaiting conversion to woodchips was evidence of the yard's role as a source of fuel, though there was no tub grinder on site. A tub grinder, which can cost up to a million dollars, is typically portable over the road system and will presumably be brought onto the site to process the logs as needed.

Arrangements for dropping off wood and arranging payment are made directly with TreeSource Solutions's buyer, Jack Santamour, who spends most of his time at TreeSource's facility in the Adirondacks and manages the Burdett yard via telephone with the help of some local employees. TreeSource is currently buying logs by the ton every Friday or by appointment.

A COOPERATIVE MODEL OF BIOMASS PRODUCTION IN DANBY

One key to sustainable local wood heat in Tompkins County is the creation of a system whereby local landowners can convert otherwise unused or underutil-

Figure 2. Entrance to the Burdett Wood Yard

Figure 3. Portable bandsaw mill at the Burdett Wood Yard

Figure 4. Wood awaiting processing by TreeSource Solutions

ized farm or pasture land to biomass production. The Danby Land Bank Cooperative (http://www.danbyland-bank.com/) provides an organization and infrastructure that allows owners of 10 or more acres in the Town of Danby to use their fields and forests (much of it marginal for farming) for grass and wood pellet production.

Built on a classic cooperative model, the goal of the Land Bank is "to unify fragmented and non-farming rural landowners to form a large enough agricultural base to provide economies of scale." Local members of the co-op lease their land to be harvested of perennial grasses as feedstocks for grass pellets or briquettes; the land is cleared for free, and the owners receive tax credits and, eventually, a share of the profits.

In operation barely a year, the DLBC has already gained 20 owner-members with more than 350 acres devoted to the project. Governance structures are in place, and plans are in the works to incorporate as a legal cooperative. The project, aided by consultation with the County Planning Department and close cooperation with Cornell Cooperative Extension, received major publicity in November with the appearance of a feature article in *Rural Cooperatives,* a publication of the U.S. Department of Agriculture (www.rurdev.usda.gov/rd/pubs/RuralCoop_NovDec09_Final.pdf).

Figure 5. First hay cutting of the Danby Land Bank Cooperative (photo courtesy of DLBC)

Establishment of a local pelletizing plant has been identified as a key to long-term sustainability and economic viability through reduction of transportation costs. The pellets, which are manufactured by grinding, drying, and extruding raw biomass into a dense, free-flowing fuel of consistent quality that can be efficiently used in inexpensive residential appliances, have a retail market value per dry ton well over twice that of the raw feedstocks. The value added more than covers manufacturing costs, so pelleting can provide an economically viable link between local biomass suppliers and the existing local pellet market.

The DLBC recently joined with Energy Independent Caroline to sponsor Town Hall meetings in Danby and Caroline regarding a company called Community Biomass Energy, which proposes to build a local biomass

pelletizing mill on Boiceville Road in Caroline just south of State Route 79. See the DBLC's newsletter (linked from their web site) for details and updates.

UNRESOLVED ISSUES

Local biomass harvesting and processing hold great promise for reestablishing the county's ability to provide for its own heating needs. However, several issues remain unresolved.

- We need to relocalize food production, too. While much of the land in the county that could produce biomass for heating is marginal for raising cultivated crops, a substantial percentage of that land could alternatively serve for rotational grazing of livestock, which is arguably a less-intensive, lower-input use of the same acreage. Thus the optimum allocation of land for biomass production vs. land for grazing or the production of winter hay remains an open question whose eventual resolution will depend on a number of variables that are difficult to predict.

- The increased use of biomass for heating will increase economic incentives to harvest wood resources beyond a level that's sustainable. The large-scale reversion of former Central New York farmland to successional forest over the last half century makes it easy to forget how quickly the forest can be cleared again. The establishment of sustainable forest management practices will be essential to the return of biomass heating as a long-range relocalization strategy.

- The rediscovery of biomass as a heat source has created a market for American wood chips as far away as Europe. Our region's potential as a major biomass producer also makes it susceptible to the kind of resource exploitation we associate with third-world countries. Heating our homes with local biomass won't succeed if higher prices cause local biomass to be exported rather than used locally.

The need for greater local control over the allocation of our local resources argues for the establishment of biomass harvesting and processing facilities under local management and provides further reason to hope for the success of initiatives such as the Danby Land Bank Cooperative and the proposed Community Biomass Energy facility in Caroline.

ONLINE WOOD HEATING RESOURCES

Cornell Cooperative Extension has posted an excellent collection of links to articles on firewood resources and heating with wood on their statewide web site at http://cce.cornell.edu/Environment/Pages/Heatingwith-Wood.aspx .

F

Funding and Finagling the Transition to Biomass Heat and Power

by Krys Cail (April 2010)

This article follows up on the previous two ("Burning Transitions" and "Heating with Biomass") with a discussion of combined heat and power applications. While continuing to focus on local efforts and local projects, the article also examines the role of local and larger-scale governmental entities in supporting the development of the biomass industry in Tompkins County and considers some roles played by local businesses and nonprofits. Some local demonstration projects that were briefly mentioned in the earlier articles are more fully considered here.

ABBOTT/LUND HANSEN LLC

The U.S., with relatively abundant biomass resources, is far behind some other countries in the use of those resources for heat and power production. This has the perverse effect of encouraging the export of US biomass resources to European countries, where both governments and businesses have embraced the development of technology and infrastructure to accommodate the use of non-fossil fuels for these purposes. Conversely, the technology needed to use North American biomass resources has often had to be imported from Europe.

In any comparison of biomass use across nations, Denmark stands out for the success it has had in weaning itself from a petroleum-dependent infrastructure. The initial motivation for this development was not an abundance of available alternative resources, but, rather, a serious brush with scarcity in the wake of the first oil shock. However, at this point, the success that Denmark has attained in maximizing efficiency in combined heat and power generation is also making Danish technology attractive elsewhere around the world. Recently, a local businessman and real estate developer and a Danish en-

gineer established a new company aimed at emulating the Danish approach to combined heat and power.

In 2010, the new company Abbott/Lund Hansen LLC was formed, joining a Danish district heating specialist with a Tompkins County developer. District heating, as a concept, is the idea of heating a number of adjacent or nearby buildings with one central heating plant. In Denmark, super-efficient heating plants may be operated on biomass fuel (pellets or chips) or traditional fuels like natural gas. Combined heat and power (CHP) is also common in the Danish systems, with the heat that is generated in the course of making electricity for a district captured and used in heating the district. Below is a synopsis of Abbott/Lund Hansen LLC's work, in the words of its founders.

Bruce Abbott and Thomas Lund Hansen recently formed a marketing and lobbying firm that is advocating for district energy in Tompkins County. A local example of district energy is at Cornell University. In 1888 Cornell built a coal fired steam heat only system for its campus. This year that system has been converted to a natural gas fired steam combined heat and power (CHP) system. Cornell's CHP system will not only supply heat to buildings on campus but it will supply 80% of Cornell's electricity needs. The only difference between the Cornell system and the systems that Abbott/Lund Hansen are advocating is that the Cornell system relies on steam and the Abbott/Lund Hansen systems relies on hot water. For the end user, hot water CHP systems are safer, more reliable, and cost less then comparable steam systems.

Combined Heat and Power systems, in general, increase energy efficiency by 30% while decreasing energy cost by 15%. There are other advantages for building CHP systems in Tompkins County. CHP systems can drastically reduce greenhouse gas emissions because they can burn a variety of fuels. For example, using biomass as fuel would reduce [greenhouse gas emissions] to virtually zero for the buildings that are connected to a biomass CHP system. Another advantage CHP systems would have in Tompkins County is that there would be numerous job opportunities building and operating these systems...

Bruce Abbott stresses that the jobs created by district generation/CHP will remain in the local economy and can't be transferred elsewhere, including the jobs harvesting and manufacturing biomass fuel. The company envisions a number of scenarios under which district generation/CHP could offer the local economy job-creation and economic development benefits. These major building projects require significant capital investment to attain a scale that can realize the efficiencies inherent in their design and reap the employment and economic development benefits. One approach that Abbott has advocated for Tompkins County is to have the AES Cayuga power plant establish and operate these districts in areas where they are practicable, such as the Downtown Ithaca Business District or the South Hill Office Campus. The new company has also suggested that Tompkins County (or the Town or City of Ithaca) might invest in the development of heating districts. The new business, Abbott/Lund Hansen, is also pursuing other opportunities to design these combined heat and power generation districts in the region; it has just signed a contract to do the preliminary design for a biomass (wood-chip) CHP system that will supply the electricity, heat, and air conditioning for 700,000 square feet of mixed use commercial and residential space in rural Pennsylvania.

It will be interesting to see what types of entities—businesses/developments, educational institutions and other nonprofits, or governmental bodies—will have the vision, the capital, and the sites to try this new approach to providing heat and power. The adoption of these highly efficient systems in the private sector can be advanced through governmental incentives to adopt the technology, which is how the Danish system came into being. What is needed is the will to transition and a plan for accomplishing the switch. Bruce Abbott puts it succinctly:

> In summary, moving toward a less costly, local, and reliable energy solution that improves energy security and environmental impact is possible today. What is required is a well-written plan and the political will to put it into practice.

CAYUGA NATURE CENTER—HEATED BY BIOMASS

Some movement exists in New York State government to subsidize the adoption of biomass heat. The New York State Energy Research and Development Authority (NYSERDA) funded a demonstration project to show how efficient and cost-effective biomass heat can be, right here in Tompkins County at the Cayuga Nature Center. The multi-fuel (woodchip or pellet) boiler used in this conversion to biomass heat was the very first unit produced by a Schenectady firm, ACT Bioenergy.[1] The firm has licensed European multi-fuel boiler technology to produce these units in New York State from all U.S.-made materials.

The 10,000 square foot Cayuga Nature Center lodge houses both educational and administrative offices for the nonprofit organization. Installation of the containerized boiler and adjacent fuel storage areas did not require any construction work or disruption of programs in the program and office space. Existing hot-water radiators were used in the retrofit, and all conversion work was kept in the basement area of the building. The three

---

1 http://www.actbioenergy.com/

Figure 1. The propane-fired system that formerly heated the 10,000 square foot Cayuga Nature Center. The system is kept on standby as a backup

existing propane boilers were kept in place to act as an emergency back-up system. The fuel and the boiler itself, in its containerized outdoor location, are an additional educational display along a path that also includes other educational exhibits and gorge overlooks used in Nature Center programs.

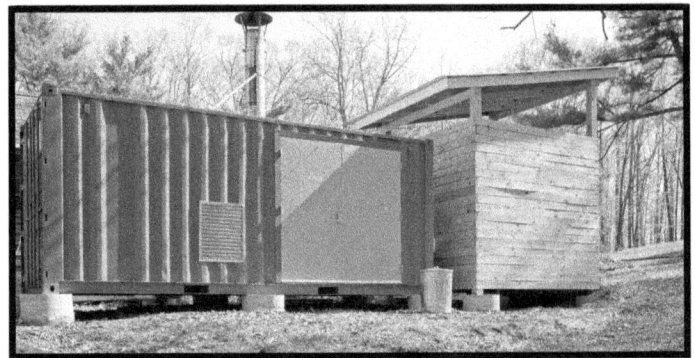
Figure 2. Exterior of new woodchip fired boiler. The wooden feed bin on the right holds about a week's worth of fuel at maximum boiler output. An auger automatically conveys fuel from the hopper to the boiler

Figure 3. The boiler can produce 400,000 BTU per hour from wood chips

Figure 4. Interior of feed bin (almost empty) showing the sweeper that moves chips across the auger trough

The program will clear a path for New York-grown fuels, create new manufacturing jobs, and improve environmental performance of biomass technologies.... ACT's project at the Cayuga Nature Center in Ithaca, NY, will demonstrate a fully automated, 90 percent efficient wood-gasification boiler technology that is proven in Europe and adapted for the U.S. market. These systems have emissions that are significantly better than conventional wood boilers and comparable to typical oil or gas boilers. Mid-sized buildings (10–100,000 sq. ft.) represent 90 percent of the boiler market in the U.S., and are prime targets for these wood systems which can achieve rapid paybacks when replacing fossil-fuel boilers.

More information on this project is available at http://www.actbioenergy.com/brochure/Cayuga%20-wood%20boiler%20photos.pdf

Figure 5. Chips are fed from below to the center of a grate at the bottom of the combustion chamber. Optimal combustion is achieved by controlling the air supplied through holes in the chip bed and holes on the sides of the combustion chamber. The ash produced by this process is less than one percent of the fuel burned

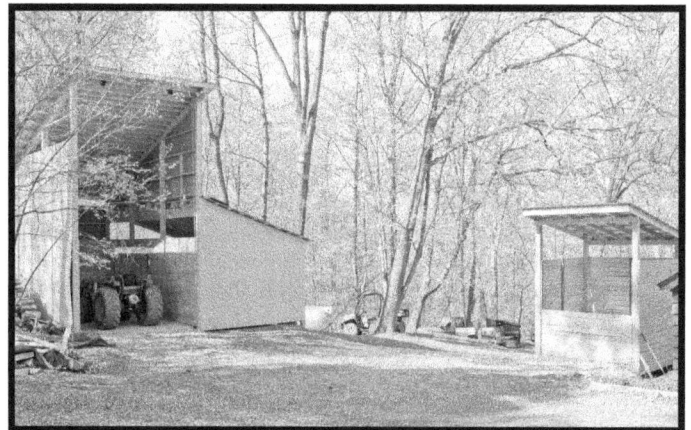

Figure 6. A 12 x 40 foot shed (on the left) stores chips to periodically replenish the feed bin (on the right). The shed was constructed with volunteer help from Cornell Engineers for a Sustainable World

This project would not have been possible without the determined and persistent effort of TC Local contributor and local biomass proponent Tony Nekut. NYSERDA was eager to have a demonstration project, and the Cayuga Nature Center was eager to solve the problem of high propane heat bills, but it took a local activist to bring the need and those with the funding together to make it work. While fuel costs have not yet been tabulated for the year, it is estimated that the new boiler will result in a 50 to 75 percent savings in fuel.

The CNC installation is part of a larger NYSERDA effort to support the evaluation and improvement of biomass-fired heating equipment. According to a recent press release,[2]

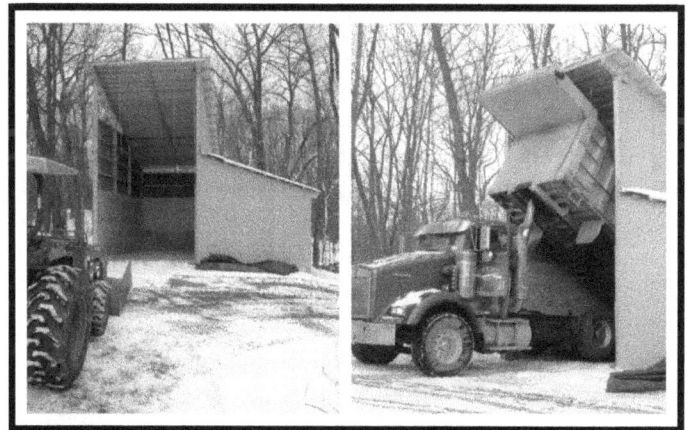

Figure 7. Left: The storage shed in winter and the front loader used to transfer chips to the feed bin; Right: Receiving a 10 ton (50 cubic yard) chip delivery from Mesa Reduction of Auburn, NY. The chips are made from the waste streams of regional lumber mills. In the future, some fuel will come from CNC and other nearby forests

---

2 http://www.actbioenergy.com/news.html#

TOWN OF DANBY HIGHWAY BARNS—PROJECT TO RETROFIT ACT BIOENERGY BOILER USING AMERICAN REINVESTMENT AND RECOVERY ACT (ARRA) FUNDS

The Town of Danby has a high level of interest in biomass as a heat and energy source. Not only are Town elected officials and staff excited about the potential of making use of a local resource in moving away from fossil fuels, the residents of the Town are also very involved. Citizen involvement is primarily through the Danby Land Bank Cooperative,[3] which "provides an organization and an infrastructure that allows rural property owners to use their fields and forests for grass and wood pellet production." In the neighboring township of Caroline, Cayuga Biomass Energy, a small group of entrepreneurs that includes TC Local contributor Tony Nekut, is attempting to start a biomass pellet manufacturing plant.

The projected cost to convert the Town's 10,000 square foot office and truck bay complex to wood chip heat is about $267,000. While the projected fuel cost savings are estimated to be 50 percent or greater, a capital improvement of that scale is difficult for a small rural township to budget or buy bonds for; usually, help from a higher level of government is needed for improvements on this scale. In this case, the Town administration decided to pursue funding under the American Reinvestment and Recovery Act (ARRA)-the federal stimulus package.

As in the Cayuga Nature Center project, biomass proponents helped to bring the need and the source of funds together—in this case, Tony and I helped the Town of Danby make application for these funds by coordinating grant-writing and project specification tasks.[4] In March of 2010, NYSERDA awarded these federal funds to Danby. For its part in the project, the Town will contribute some highway worker hours to the excavation and concrete work needed to construct a covered fuel storage area. The boiler unit, which is almost identical to the one in use at Cayuga Nature Center, will be installed by a regional heating contractor, and the jobs producing biomass fuel will be hyper-local—ideally, in Danby or adjoining Caroline. In fact, the Town Highway crews plan to produce some of the wood chip fuel themselves in the process of keeping the roadways clear. This is a good use of a federal program aimed at maintaining and creating jobs in economically distressed counties like Tompkins.

RPM ECOSYSTEM'S COMBINED HEAT AND POWER PROJECT/BIOMASS DEMO PLANTATIONS

PJ Marshall, one of the principals of RPM Ecosystems,[5] wanted to provide the heat and power to operate the firm's Town of Dryden greenhouses and company headquarters while remaining carbon-neutral. And she wanted to do so using only the products RPM grows—native hardwood trees. Additionally, she sought to develop and demonstrate a biomass plantation system using native hardwood trees planted specifically for a combination fuel/lumber harvest, staged to produce first fuel wood and then lumber, over a number of years, while maximizing forest canopy and carbon sequestration throughout the process. RPM pursued this plan through local Congressman Michael Arcuri, looking to secure a federal appropriation to fund the project.

The company made good progress in developing the project and getting the appropriation drafted last year (2009) but then encountered difficulties when Congress passed a rule requiring that no appropriations go directly to private companies. RPM regrouped and engaged TCAD[6] as a fiscal sponsor for the projects. Heather Filiberto, Director of Economic Development Services at TCAD, describes the agency and its role in the project this way:

> TCAD, the County's lead economic development agency, is a non-profit organization whose mission is to build a thriving and sustainable economy that improves the quality of life in Tompkins County by fostering the growth of business and employment. In situations in which governmental funding must be received by a non-profit, TCAD has stepped in and sponsored applications on behalf of local entrepreneurs in the past. TCAD has agreed to sponsor this request for federal funding on behalf of RPM.

In order to succeed in getting an appropriation in the federal budget for a project, the applicants must obtain letters of support from a wide variety of local officials. The typical support letter is prepared by the applicant in overall substance, then transferred to letterhead and signed by the various elected officials with only slight modifications. The projects are briefly described along with the expected benefit to the community. The following excerpt, from Senator James Seward's letter, demonstrates the approach.

> I am writing to express my strong support for Tompkins County Area Development and RPM Ecosystems Ithaca LLC's, innovative Dryden, New York, green building and renewable energy project titled

---

3 http://www.danbylandbank.com/site/home.html

4 Contact Tony Nekut or Krys Cail through the comments section linked to the online version of this article at tclocal.org if your Tompkins County municipality or school district is interested in pursuing biomass heat funding; we are interested in sharing information.

---

5 http://www.rpmecosystems.com/

6 http://www.tcad.org/

Distributive Biomass Combined Heat and Power for CO2-Neutral Facility Operations....

...this project helps install and commission a 200KWe distributive biomass combined heat and power set for sustainable/renewable electricity and thermal energy production in support of RPM Ecosystems Ithaca LLC's operations....

TCAD, RPM Ecosystems, and Congressman Arcuri are all hopeful that the funding for this project will be included in this year's federal budget. Still, the project must wait to commence until the political process runs its course.

INDIVIDUAL HOMEOWNERS CAN ACCESS GOVERNMENTAL BIOMASS INCENTIVES

Some government-assisted financing options exist for individual homeowners interested in converting some or all of the heat or hot water produced in their homes to biomass fuels. Anyone who is in a position to benefit from a tax incentive can receive up to 30 percent of the cost of a pellet stove (not to exceed $1,500) in tax savings. A website is available to help with determining whether this program meets your needs,[7] or contact the Pellet Fuels Institute.[8] Local pellet stove merchants can also assist in understanding the program and which units qualify. Unfortunately, stoves and furnaces that burn cordwood are not eligible for these incentives.

NYSERDA also has some homeowner financing programs[9] for the installation of a pellet stove and for the energy efficiency retrofits that can be accomplished in conjunction with a transition to a heat source based on certain kinds of renewable fuel. In general, cordwood stoves and furnaces are ineligible for these programs. For homeowners with low or moderate income, low-interest financing programs, and even some grants, are available through Ithaca Neighborhood Housing Services.[10] Similar programs are available through Tompkins Community Action,[11] and some similar services may be available through Better Housing for Tompkins County[12] as a part of home rehabilitation. All of these housing agencies should be contacted to determine what programs might work best for your individual needs.

Most programs will require that you obtain a professional energy audit to determine which energy improvements may be most cost-effective for you. Even if you don't use an incentive program, an energy audit can help you to tackle energy investments in the order that gives you the most benefit for the money invested. Conservation measures and efficiency upgrades are often more cost-effective than investing in a renewable fuel heat source. The housing agencies linked above can provide referrals for homeowners of all incomes to qualified energy audit providers and Building Performance Institute (BPI) certified contractors. In most cases, only BPI-certified contractors are eligible to perform work that will qualify for incentives. These energy auditors and BPI-certified contractors are also trained to make use of up-to-date methods and products for saving energy and using renewable fuels.

FUNDING AND FINAGLING: NEGOTIATING THE POLITICAL PROCESS TO TRANSITION TO BIOMASS

Local, state, and federal governments are involved in energy policy and the implementation of energy projects in a number of different and evolving ways. Even a very savvy and motivated community such as Tompkins County may find it difficult to work the system well enough to get sufficient funding and financing for transitions to carbon-neutral and renewable fuel sources. Over time, government-funded energy efforts at conservation, which should always be the first step in a sustainable energy plan, have become institutionalized in a way that makes them more accessible to homeowners, businesses, and other community institutions. However, renewable energy conversions remain new enough that the path to government sponsorship is not always clear —in both the sense of "visible" and "free of obstructions."

Some motivated activists claim that the slow grinding of the gears in the public sector is not worth the patience to accommodate. The fastest and best approach when projects are low-tech and inexpensive may be a community barn-raising kind of effort. However, commercial-scale projects in large buildings, or the highly efficient district heat and power systems that group many buildings in a densely developed area on one heating system, can't easily be accomplished via small-scale community efforts. Both funding and implementation will typically require some level of governmental assist or substantial private investment of capital.

How do thinkers, planners, and activists work most effectively to bring about a transition away from fossil fuel dependence? Understanding the ways that the layers of government divvy up responsibility, and how they do and don't collaborate, is an important place to start when developing a strategy.

Planning efforts go on at all levels of government— federal, state, regional, county, and municipal. Professional planners are often those who elected officials turn to for information and explanation of policy options, even though elected officials themselves enact policy. It

---

7   http://energytaxincentives.org/consumers/heating-cooling.php
8   http://www.pelletheat.org/3/residential/taxCredit.html
9   http://www.getenergysmart.org/SingleFamilyHomes/ExistingBuilding/HomeOwner/Financing.aspx
10 http://www.ithacanhs.org/pdf/LendingServicesWeb020210.pdf
11 http://www.tcaction.org/energy.htm
12 http://www.betterhousingtc.org/bet2_rehab.html

is productive to educate both planners and elected officials about new policies on renewable energy enacted by other governments and to call their attention to demonstrations of new technology. By definition, planners are charged with taking the long view of our situation, and may be the first to show interest in emerging technology and trends. Eventually, however, elected officials must choose to implement new projects.

Those of us, planners or otherwise, who take a long view of our local adjustment to energy descent may consider funding for transitions away from dependence on fossil fuel to be one of the most vital things our governments can do to assure our future security. Implementing that transition can be accomplished by educating elected officials and the professional planners who advise them, and also by applying for and using the funds (grants and capital) and financing (low-interest loans and tax-exempt bonds) for the purpose when such are available. The process is likely to be difficult, even frustrating at times. To lead the way to a renewable-fuels future, we should focus on creating the will, knowledge, and capacity to make good use of every opportunity for implementing projects. The more we show each other how to heat with renewable fuels, the more examples of successful projects will be available to help others understand the benefits. Eventually, we will reach a tipping point at which the logic of using sustainable, renewable sources for our heat and power will make more sense than fighting one another for a rapidly-diminishing stock of polluting fossil fuels.

R

Relocalizing Investment in Our Local Food System

By Krys Cail (June 2011)

Over the past decade or so, social trends have emerged that promote local economic exchange around a regional or local food system. The rise in popularity of farmers' markets is shown by the 16 percent increase in number of markets between 2009 and 2010.[1] Grocery stores, college cafeterias, and now even Walmart stores are trying to source more fruits and vegetables from local growers.

Slow Food is an international NGO that began in Italy but is now world-wide in scope. It celebrates the local and regional culture of the table while encouraging taking the time to enjoy basic social activities, like sharing food. Indirectly, the Slow Food movement also encourages local culinary, agricultural, and wine tourism industries. In the Tompkins County area, we have made significant progress in the development of local and regional food systems, often in collaboration with the region's grape growers and wine makers. Local food is a current focus of local interest that we would be well served to further develop in view of energy decline and the need to shorten food supply chains.

The Slow Food Movement was among the inspirations for the work of Woody Tasch, a socially responsible investing leader and author. He coined the term "Slow Money" to describe investing in the local foodshed with a portion of one's portfolio—with an understanding that this investment might pay off better in social and environmental benefits while generating a somewhat lower financial return. His book, *Inquiries into Slow Money: Investing As If Food, Farms and Fertility Mattered*,[2] inspired others, and a number of like-minded individuals launched an effort aimed at starting a Slow Money Movement.[3] They adopted a goal—one million people investing one percent of their assets in local food systems within ten years. They also adopted principles[4] and began working with local and regional Slow Money organizations to establish investment programs. Slow Money has gained some national recognition over the past couple of years, with articles appearing in *Business Week*[5] (one of their "big ideas for 2010"), Entrepreneur.com[6] (one of "five financing trends for 2011"), *Utne Reader*,[7] *Time*,[8] the *Wall Street Journal*,[9] and the *Los Angeles Times*.[10]

A local group, loosely affiliated with the national movement, has begun planning activities here in Tompkins County. This Slow Money Central New York group can be contacted through the Alternatives Business CENTS program[11] or Local First Ithaca.[12]

Envisioning a new investment paradigm is difficult theoretical work, but actually implementing a system that directs flows of investment cash into local food systems is even more difficult. As a nascent movement, Slow Money has moved methodically to build a robust infrastructure for implementation. A growing national network of interested people has been considering how local groups or "Slow Money Alliances" would be structured in order to accomplish the work of bringing more investment into local food systems. The national Slow Money Alliance uses a number of other national organizations as models, including Slow Food, BALLE (Business Alliance for Local Living Economies), Social Ventures Partners, and Transition US. There is a focus on preparing for energy descent through relocalization by investing in local food systems.

Investors may have a simple need—to keep at least a portion of their portfolio invested in the local foodshed. Food buyers, both in the urban areas in the region and in the Tompkins County area, also want to buy food from nearby. This gets complicated very quickly, however, by rural/urban interdependence. Cities, and especially huge port cities like NYC, relocalize by becoming more dependent on a regional, not local, foodshed. Rural areas in the region may be dependent on investment from the urban areas. Tompkins County may or may not be in a position in the future to source investment

1 http://www.ams.usda.gov/AMSv1.0/ams.fetchTemplateData.do?template=TemplateS&leftNav=WholesaleandFarmersMarkets&page=WFMFarmersMarketGrowth&description=Farmers%20Market%20Growth&acct=frmrdirmkt

2 http://www.slowmoney.org/book.html

3 http://www.slowmoney.org/

4 http://www.slowmoney.org/uploads/1/3/6/7/1367341/principles.pdf

5 http://www.businessweek.com/smallbiz/running_small_business/archives/2009/12/big_ideas_for_2.html

6 http://www.entrepreneur.com/article/217795

7 http://www.utne.com/Politics/Utne-Reader-Visionaries-Woody-Tasch-Slow-Money-Alliance.aspx

8 http://www.time.com/time/business/article/0,8599,1921889,00.html

9 http://online.wsj.com/article/SB125305092106313571.html

10 http://articles.latimes.com/2009/sep/22/business/fi-smallbiz22

11 http://www.alternatives.org/cents.html

12 http://localfirstithaca.org/

capital from local investors alone; rural areas may find that they continue to have some dependence on larger regional centers of finance. Many Tompkins County farm and food businesses currently sell a portion of their produce to local markets, and also ship a portion to regional urban population centers, most typically NYC. Tompkins County is within the NYC Greenmarkets catchment area, and currently, many local food producers make the trip to sell in those lucrative markets. While that pattern may change some as the price of truck transportation increases markedly, it may not: sourcing fresh foods from even farther away may cost yet more, making the relative cost of Tompkins County grown food in NYC still attractive.

Additionally, wholesale foodstuff supply chains move food from local farms and food processors into urban markets. Locally-owned shipping companies, such as Regional Access,[13] may adapt to new transportation approaches as fossil fuels increase in price. For example, multi-modal shipping via train and/or barge would allow shelf-stable or cooled produce to travel more economically. One gallon of fuel will take a ton of freight about 155 miles by truck, 413 miles by train, and 576 miles by barge.[14] In particular, crops such as grains, beans, seeds, oils, and meats that require a large land base for their production are likely to continue to be imported into large cities from their peripheral rural areas. In many cases, it's more cost-effective to manufacture minimally processed foods, such as canned or dried fruits and vegetables, closer to where they are grown, and then ship them via lower-energy transport, such as barges or trains. Tompkins County is exceptionally well placed to ship local goods by water; it is possible to send goods by boat from Ithaca to anywhere on the Great Lakes, the Mississippi, or the East Coast.

The Central NY Slow Money Group has been meeting at the Alternatives Federal Credit Union.[15] The group has established a cooperative, interdependent relationship with Slow Money NYC. Central NY generally, and Tompkins County in particular, has many farm and food enterprises, but relatively fewer eager high-net-worth investors. For NYC, that situation is reversed. Some collaboration can be of value, allowing people who eat Tompkins County food to invest in Tompkins County food growers and processors, whether they live very near the farm or in the nearest megalopolis.

Access to capital can be gained by a business through an equity deal (selling portions or shares of business ownership) or through debt instruments (loans requiring a stipulated repayment schedule, but conferring no ownership rights). There are also hybrid arrangements,

such as debt instruments that convert to equity shares if not repaid over a certain period. Under the current regulatory framework, it is difficult—not impossible, just difficult—to raise private equity funds for a business venture from a large number of investors of limited means. Typically, "qualified investors" (those with more than one million dollars in net worth) are able to play by somewhat different rules than the rest of us, as the regulators consider them to be savvy enough to fend for themselves in the investment world. To make an offering to a group of people who are not all "qualified investors" (for instance, the membership of Greenstar Cooperative Market), some form of an intermediary fund is probably most practical.

Cooperative membership/ownership organizations are but one model that allows for a large group of investors to provide capital and share risk. The CSA (Community Supported Agriculture) model is another approach. Slow Money groups at the national, local, and NYC levels are all exploring the best ways to facilitate these kinds of transactions, meeting the needs of businesses while mitigating exposure to risk for investors and also keeping some liquidity for investors.

Slow Money group members seek to meet two very different kinds of needs with one suite of mechanisms.

First, investors want to move beyond socially-responsible investment opportunities and now want to invest their money in businesses that have a triple-bottom-line benefit: businesses that are socially responsible and environmentally appropriate while also making some profit. People who understand the inevitability of energy decline may well want their money invested in shortening the supply chains for essentials like foodstuffs.

Second, small farm and food businesses need access to capital to grow and process the foodstuff supplies that we need in a more localized or regionalized food system. Traditional financing, still stuck in a global market worldview, is often disinclined to channel investment into the type of enterprise that could help smooth the adjustment to a world with a lot less oil.

THREE OPPORTUNITIES FOR INVESTMENT IN TOMPKINS COUNTY FOOD SYSTEMS

Several local initiatives offer both Tompkins County residents and city dwellers the opportunity to invest "slow money" here. In the following, I'll briefly describe three of them.

Local opportunity number 1: Facilitating land acquisition by prospective farmers trained at Groundswell

Groundswell[16] is a program that uses both classroom teaching and on-farm training to teach students to farm.

13 http://www.regionalaccess.net/Home.html
14 http://www.waterwayscouncil.org/study/public%20study.pdf
15 http://alternatives.org/
16 http://www.groundswellcenter.org/

If there is one practical suggestion for an easier transition in the face of energy decline, it is that more people need to learn to be able to grow food. In a globalized market for energy, food-growing resources have been diverted to the production of fuels such as ethanol, which, in combination with increases in costs of petrochemical inputs into industrial farming, has caused food commodities to experience great price volatility. In the near term, we are likely to see spot price run-ups and shortages, while in the long run, food grown closer to home and with more animal-power, human attention and labor, and organic inputs will be more sustainable. Groundswell's programs are tailored to producing more farmers through a classroom-based curriculum of instruction delivered at EcoVillage at Ithaca, dovetailed with hands-on farming apprenticeship in a structured program that exposes students to many local farms. The emphasis on sustainable and organic methods prepares new farmers to farm with less reliance on fossil fuels. When newly-trained would-be farmers emerge from this training, however, they require land to farm.

Some communities, such as Burlington, Vermont, have established agricultural land specifically set aside for use by beginning farmers. The Intervale in Burlington is an area that includes community gardens as well as small acreages for use by tenant farmers who are just starting out in vegetable farming. The land has excellent soil and is close to housing in the city. The location is ideally suited for this purpose, and the property has been protected from development by the generous action of a philanthropist. Tompkins County currently lacks such tenant-farming options, but Groundswell is attempting to develop similar options locally.

Joanna Greene, Executive Director of Groundswell, has worked with local farmers and EcoVillage to establish a farm incubator program in Tompkins County. If an intermediary financial capital stream were available, the graduates of such programs would be ideally suited to match with a group of local investors. Alternatively, CSA models or direct equity investment on the part of larger, qualified investors, or debt-to-equity financing, may be more appropriate financing approaches. A "Slow Money" program could take a number of forms. Joanna has been participating in Slow Money planning talks, representing the needs of beginning farmers.

Local opportunity number 2: Facilitating grain processing for local grain farmers through Farmer Ground Flour

Grain farmers Erick Smith and Thor Oeschner joined forces with Greg Mol about a year ago to begin a grain-milling operation in Trumansburg, Farmer Ground Flour.[17] They use a modern mill that can make up to 15,000 pounds of flour a month. They began by grinding the wheat, spelt, corn, rye, and other grains they grew. At first, they had to take the grain to Penn Yan to dehusk it, but now that operation is handled at one of the farms.

They clearly hit an area of the food system ready for development. By January 2011 they were in the *New York Times* in an article titled "Reviving New York State's Grain Belt."[18] To quote from the article:

> It is a cooperative effort among several farms growing organic corn, spelt and wheat, often heirloom varieties…. Packaged under the Farmer Ground Flour label, the flours are sold in paper sacks in Greenmarkets by Cayuga Pure Organics, a participant in the cooperative. The flours are fresh, and have not sat for months in warehouses.

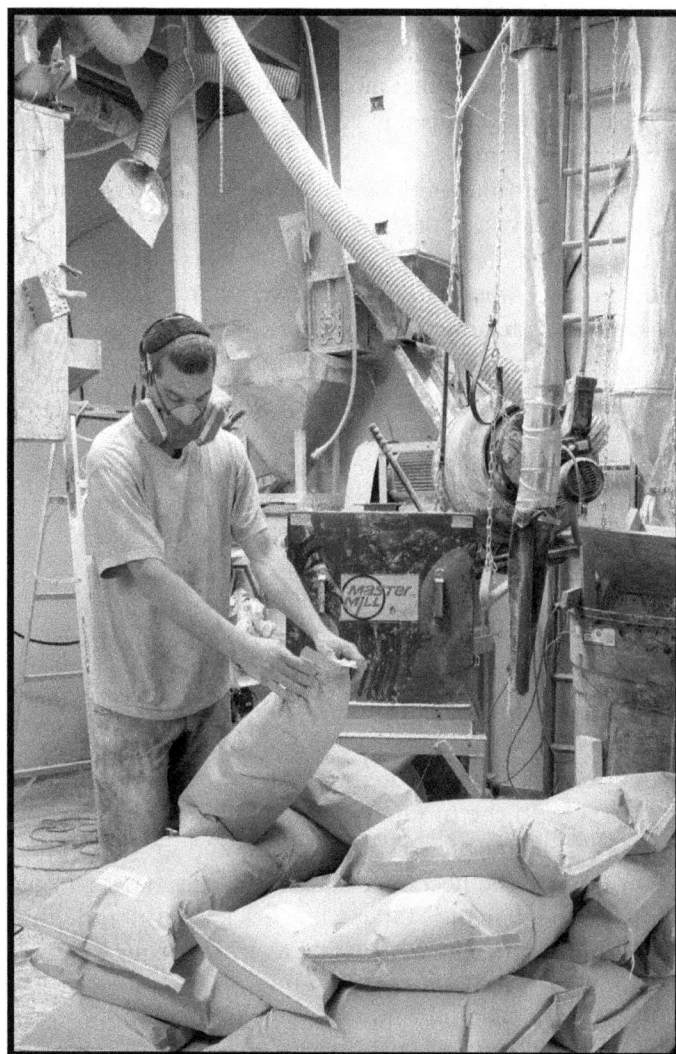

Figure 1. Greg Mol bags rye flour at Farmer Ground Flour

Located at the old Agway building in Trumansburg, the mill occupies a site where animal feed was milled in the past. In many ways, it is an ideal site for an enterprise that includes a lot of unloading grain from trucks and a

17 http://farmergroundflour.squarespace.com/

18 http://www.nytimes.com/2010/01/06/dining/06flour.html

lot of loading flour and other milled products back onto them.

Farmer Ground Flour has been very successful in meeting an emerging need for artisan-milled flours and meals in NYC. That is not, however, their only emphasis. They also sell flour and other milled products to the local market, through Regional Access, Greenstar Coop Natural Foods Market, and Garden Gate home delivery service. A recent edition of *GreenLeaf*,[19] the newsletter of Greenstar Coop Natural Foods Market, showed the big-picture development of Farmer Ground Flour in historical perspective:

> Farmer Ground's success is part of a larger effort to restore grain growing to New York state. While now thought of as dairy country, upstate New York once grew so much grain that Rochester topped the nation's flour production in the mid 1830s, giving it the nickname "Flour City." (A later rise of nursery businesses changed that moniker to "Flower City.") That flour was shipped to New York City and beyond via the Erie Canal.

> Oechsner and Smith have both worked closely with Elizabeth Dyck, of the Organic Research and Information Sharing Network, which seeks to reintroduce wheat growing to New York state… [She] is working with farmers like Oechsner to identify those [varieties] that grow well in New York's challenging climate, and, just as importantly, also taste great and bake well.

> …Like other foods, "the flavor has been bred out of wheat," [Oeschner] explained, in favor of yield and uniformity. "Growing the old wheat varieties is like growing an heirloom tomato."

> "Farmer Ground Flour is really making a difference for other farmers," said Dyck. "They're a great example of farmers banding together to put needed infrastructure into place, in this case a milling facility. They deserve enormous amounts of respect."

No question, there is market interest, both regionally and locally, in the product of a local grain mill. But how does a small, "farmer-owned, grown, and ground" operation finance the necessary equipment purchases to keep up with the demand? Greg Mol, Erick Smith, and Thor Oeschner approached banks to seek financing for their equipment needs, but the amount of money that they sought to borrow was too small to fit the lending programs available. They have pursued working with individuals in the community to finance their equipment needs, but there is no organized program for doing so. The need for relatively small infusions of capital hampers their ability to expand and improve Farmer Ground Flour.

Will grain farming for human food expand in New York State only as quickly as the processing capacity is able to expand? Slow Money could be a means by which those interested in the re-development of grain farming in New York State could participate in the effort to develop the needed processing capacity. Greg, Erick, and Thor have already made connections with a few local investors to gain some access to expansion capital and hope to do more of this in future. And the availability of their product has already spurred other business start-ups and more local investment opportunities. For instance, Wide Awake Bakery[20] operates a bread CSA using Farmer Ground Flour as an input.

Local opportunity number 3: Expanding Cayuga Pure Organics into rolled grains

Erick Smith is not only a partner in Farmer Ground Flour, he is also a principal of Cayuga Pure Organics.[21] Cayuga Pure Organics is the source of much of the "locally grown" beans and grains offered for sale in the Greenmarkets, co-ops, and restaurants of NYC. They also supply our local Tompkins County region with these products. Cayuga Pure Organics was also featured in a *New York Times* article this year, in the Magazine under "Field Report — Market Watch."[22] This excerpt shows how Cayuga Pure Organics evolved to serve the niche market it now depends on, growing grains and beans for human consumption, to be sold in Tompkins County and NYC:

> In 2003, Erick Smith and Dan Lathwell—men nearing 60 who'd farmed intermittently when not working at Cornell or teaching elsewhere—thought they'd hit upon a smart niche when they created Cayuga Pure Organics to grow pesticide-free feed for the region's newly organic dairy farms. Two years later, the Ithaca food co-op and a natural-food distributor asked if they'd grow organic beans on their land in the town of Caroline. They were also connected with a local taqueria, and soon the two were struggling to keep up with the restaurant's weekly order for 500 pounds of black and pinto beans. Then, in the fall of 2008, the farm inspector for New York's Greenmarket tracked them down in her quest to find a grower to satisfy the demand for local beans and grains. "We hemmed and hawed, thinking that going to New York City is a whole step up in the organizational process," said Smith, an articulate man for whom overalls and a graying beard are a natural fit after years of teaching math education. It also required getting up to speed in marketing, which for farmers means both self-promotion and literally selling at markets.

19 http://www.greenstar.coop/index.php?
   option=com_content&task=view&id=565&Itemid=219

20 http://www.wideawakebakery.com/
21 http://www.cporganics.com/live/
22 http://www.nytimes.com/2010/10/17/magazine/17food-t-000.html

Figure 2. Owner Erick Smith with wheat cleaning equipment at Cayuga Pure Organics

Many of the processes for harvesting, shelling, and cleaning the beans and grains can be handled directly on-farm. Over time, Cayuga Pure Organics has become less dependent on other farmers for the use of processing equipment, streamlining the efficiency of the operation. However, specialized equipment for such tasks can be expensive, and it can be difficult to raise the capital needed to purchase it, house it, and integrate it into the operation. For some time, Cayuga Pure Organics has had plans to purchase equipment to be able to roll oats. Oats are a crop well-suited to our climate in Tompkins County, but they are almost always consumed by humans in the form of rolled oats, also known as oatmeal. Cayuga Pure Organics applied to the NYC Slow Money Group's first Entrepreneurs Showcase to pitch the idea of investing in this business expansion. They were one of only ten businesses that will be featured in the first Showcase, giving them the opportunity to gain Slow Money investment for this business expansion.

CONCLUSION

The area between the growing consciousness on the part of consumers that they want to support a more localized food chain on the one hand and farmers who want to grow and provide local foods on the other is ripe with possibility to re-invent investment. While the shape of this emerging movement is not yet clear, the motivations of farmers, food processors, short-haul food transporters, and restaurant chefs are clearly aligned with those of investors with an interest in facilitating a more localized farm and food sector. The roles of regional investors, and the roles of local investors, will be established in part based on who steps forward to help shape the food web through investment and marketing. Perhaps, depending on developments, the Ithaca Hours local currency revival will also play a role. Establishing a farm and food sector in Tompkins County that is able to provide grains, beans, oils, meats, and dairy products to the metropolitan areas of the region as well as the local market seems a relocalizing strategy worthy of the investment of both thought and money.

POSTSCRIPT FROM THE FARMER: ERICK SMITH NOTES SOME ADDITIONAL BENEFITS AND HURDLES

Erick Smith of Cayuga Pure Organics read an early draft of this article and responded with the following note, which he has kindly given us permission to include here.

The basic products we produce are helping support others in the community. Farmer Ground Flour is one such startup. Another is Wide Awake Bakery in Mecklenburg... Ron Springer in Van Etten is using our grains to produce sprouted products including sprouted gain crackers, sprouted rolled grains, and sprouted breads. Also, Hans Butler, an Ithaca-based chef, is actively developing products from our beans and grains under the name *Cayuga Pure Organics, Chef Hans*. He is currently producing the bean dips that are available at Greenstar and is in the process of developing other products. This year we are also growing mustard seed for Mary Graham, a local mustard producer. The point is that Cayuga Pure Organics and Oeschner Farms, as producers of basic organic commodities, provide the basis for other small-scale food processors to create their own products based on our locally-grown commodities.

A major struggle that both CPO and Farmer Ground Flour face is the lack of infrastructure to support what we are trying to do. 100 years ago, operations like ours were scattered across NY State and there was appropriate equipment, repair parts, local expertise, and market structures in place to support these operations. One of our technical and financial challenges is recreating this infrastructure in a modern world where few models are available.

Another major issue we both face is that, compared to conventional farms and conventional flour mills, we are very, very small-scale, yet from the perspective of many of the producers of local produce, we seem large. A major reason is that growing grains and beans and milling flour require a certain level of mechanization that forces certain economies of scale. So the tractors we use are small compared to what would be found on typical crop farms, and the 40-year old combines we use for harvest are so small that the size machine we use is no longer even available new. To operate on a smaller scale would make our products prohibitively expensive. Yet, because we are so mechanized, we are very dependent on fossil fuel energy. Farmer Ground depends on electricity and the farms depend on diesel fuel. We know that this has to change and that we face a major challenge in creating that change. Greg is currently actively looking into the prospect of using water power to produce the electricity to run the mill. We currently are using about 10% bio-diesel and would like to use more, but the older diesel engines in our equipment can have problems with higher levels of bio-diesel. So, we know change is coming and may very well, at some point, be looking for ways for the community to support our efforts, both technically and financially. If the Slow-Food and Slow-Money communities are serious about supporting the needed changes for a local-foods economy, these are issues that we all need to be looking at together.

About the Authors

Jon Bosak, TCLocal Editor, chairs an international technical standards committee and is a member of the Town of Ithaca Planning Board.

Krys Cail is a consultant working with small businesses and government on local and regional business and economic development planning, food systems, and sustainable agriculture. She is the convener of the local Slow Money Central NY group and chair of the NOFA-NY Gas Drilling Subcommittee.

Tony Nekut was a successful local inventor who contributed much to the local sustainability movement in projects to produce and use local biomass for energy production. We are the poorer for his passing.

Karl North experimented with models of agricultural sustainability for 30 years as owner-operator of Northland Sheep Dairy, taught Ecological Agriculture at SUNY Binghamton, and is currently managing construction of his second attempt at passive solar residential design.

Bethany Schroeder, TCLocal Secretary, is a local writer and healthcare worker and Executive Director of the Ithaca Health Alliance.

Tom Shelley, TCLocal Chair, is a Sustainable Tompkins director and is active in many other aspects of the local sustainability movement.